HYDRO

THE DECLINE AND FALL OF ONTARIO'S ELECTRIC EMPIRE

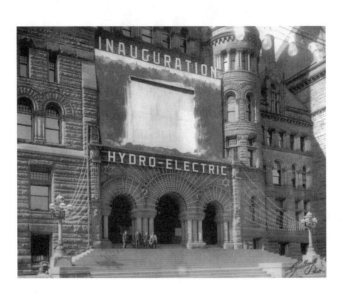

HYDRO

THE DECLINE AND FALL OF ONTARIO'S ELECTRIC EMPIRE

Jamie Swift & Keith Stewart

Between the Lines
Toronto, Canada

Hydro

First published in Canada in 2004 by
Between the Lines
720 Bathurst Street, Suite 404
Toronto, Ontario M5S 2R4
1-800-718-7201
www.btlbooks.com

Library and Archives Canada Cataloguing in Publication

Swift, Jamie, 1951–
        Hydro : the decline and fall of Ontario's electric empire / Jamie Swift & Keith Stewart.

Includes bibliographical references and index.
ISBN 1-896357-88-1

        1. Ontario Hydro—History. 2. Electric utilities—Government
ownership—Ontario—History. I. Stewart, Keith, 1967– II. Title.

HD9685.C34O5 2004   333.79'32'060713   C2004-904135-5

Front cover image: csaimages.com
Frontispiece: *Ontario Hydro Inauguration May 2, 1911*; photo by Alexander Galbraith/Galbraith
Collection—The Globe & Mail Photo Store
Cover and text design by Jennifer Tiberio
Printed in Canada by union labour

Between the Lines gratefully acknowledges assistance for its publishing activities from the
Canada Council for the Arts, the Ontario Arts Council, the Government of Ontario through the
Ontario Book Publishers Tax Credit program and through the Ontario Book Initiative, and the
Government of Canada through the Book Publishing Industry Development Program.

Canada Council   Conseil des Arts   Canada   ONTARIO ARTS COUNCIL
for the Arts   du Canada   CONSEIL DES ARTS DE L'ONTARIO

Contents

Preface   ix

Acknowledgements   xi

one   Introduction: The Rise and Fall of an
Electric Empire   1

two   Power and the Tory Dynasty: The Making
of a Crisis   8

three   The Electric City and Public Power:
Promise, Peril, and Peterson   32

four   Liberalizing Electricity: Chile, Britain,
and Ontario   57

five   The NDP Years: Clearing the Tracks
for Privatization?   73

six   South of the Border: California, Enron, and
the Yankee Alternative   96

seven   All Aboard the Privatization Express:
The Harris Tories   117

eight   Danger Ahead: The Wheels Fall Off   145

nine   Morbid Symptoms and a Regime Change   184

A Chronology of Ontario's Electricity System
1884–2003   213

Notes   217

Index   233

Preface

ONTARIO HYDRO WAS, for nearly a century, one of the most successful public utilities in the world. The company always hired talented people to undertake its ambitious ventures. Its official historian was Merrill Denison, English Canada's first important twentieth-century playwright and a man who did not shy away from metaphorical claims that reflected the ever-confident company's self-image. Denison wrote that the northern land's rocky plateau forced Canada to remain a have-not country until Hydro deployed science and technology to unlock nature's bounty: "It is the land that provides the people with their economic sustenance, molds their character, determines their social and political outlook." Hydro's mythmaker may have been borrowing from Harold Innis—and anticipating Chairman Mao—when he stated in 1960 that, in tandem with the internal combustion engine, Hydro had wrought a "beneficent cultural revolution."

The myth endured. If you make your way to the stacks of a large library in Ontario and search out the shelf where they keep the books on Ontario Hydro, you will notice that most of the titles play on a single word. *The People's Power* (Denison's book). *The Politics of Power*. *Delusions of Power*. *Public Power*. *Power at Cost*. *Power at What Cost?* There is at least one case, too, in which the writer and publisher did not choose a wordplay on the intimate, even incestuous, relationship between electrical energy and political and economic power in the province. That book bears the simple but suggestive title *Electric Empire*.

This book spans twenty-five years of political and energy history in Canada's largest province. It begins with a gathering of alternative energy activists protesting the construction of a huge nuclear power plant. It ends with the ignominious collapse of what one of our informants described as "a very expensive experiment in neo-classical economics." The intervening period was marked by the demise of the apparently omnipotent institution that, as much as any other, had shaped modern Ontario.

The progression from power (in all its senses) to empire and later to decline and fall is the subject of this book. How could an electric empire that had long been so very powerful fall with so little public controversy? And how did the first attempt to replace it go so very terribly wrong? This is the story of how a swashbuckling company met its demise. We sketch the churn of politics, economics, and ideology, all of which played a part in the decline of Ontario's electric empire.

We came to the task from different backgrounds but with a shared perspective. One of us has, over the course of ten books and years of journalistic experience, become ever more convinced of the need for what George Orwell called a political purpose in writing—using the word *political* in its widest sense: "Desire to push the world in a certain direction, to alter other people's idea of the kind of society that they should strive after." The other has had an academic interest in political ecology and a long-standing commitment to the environmental movement. But green activism has been frustrating to the extent that it has required distilling complicated issues into easily digestible sound bites. Hence the need to produce a volume that, we hope, will inform a democratic debate and help strive after an electricity system that will not poison the planet.

What follows reveals our sympathy with the people who gathered at Darlington in 1979. They warned that nuclear power was an expensive, dangerous folly, predicting nothing but grief from Ontario Hydro's commitment to a nuclear future. But even though their concerns have since been validated, we should not succumb to a smug "We told you so." That would contribute little to an understanding of the decline of the electric empire. More importantly, it would not help at all in reflecting how this little slice of history might help us to imagine ways of avoiding the mistakes of the past.

Acknowledgements

We thank Janet Pearse and Erica Wilson, whose patience with the process of our research and writing is matched by their lovely and lively company. Also, everyone who agreed to be interviewed for this book. Special thanks to Charlene Mueller and Paul Kahnert, for doing what they did, and to CEP and CUPE for doing what they did. Also the staff, board, and volunteers at the Toronto Environmental Alliance and the staff and volunteers at Between the Lines. And especially to Robert Clarke, that steadfast anchor of the battery. No wild pitches get past him.

Jamie Swift
Keith Stewart

one

Introduction: The Rise and Fall of an Electric Empire

Hydro was nationalized in such a way and then managed in such a manner as to debase the concept of public ownership and discourage the extension of the principle. . . . Public ownership is in itself a neutral phenomenon. Its origins, benefactors and behaviour determine its character.
— H.V. Nelles, *The Politics of Development*, 1974

FOR TWENTY-FOUR YEARS Allan Kupcis worked for what was once one of the largest utilities in the world. The son of a Latvian stonemason, Kupcis got a job at Ontario Hydro just after completing his postdoctoral studies at Oxford University, where he did research on the physics of copper single crystals. It was 1973, and if you were interested in matters nuclear, Canada's biggest Crown corporation was the place to be. In Ontario, where the concession roads had been successfully wired up back when most U.S. farmers were still using kerosene lamps, the company was known simply—and affectionately—as "Hydro."

The star physicist started out in research, supervising a team of engineers at the huge new Pickering A generating station just east of Toronto on the Lake Ontario shoreline. The team was searching for ways of doing remote inspections of the in-reactor components, looking into how the company could investigate the guts of the reactors without exposing workers to lethal doses of radiation. It was the best of times to be at Hydro. The giant company was self-confident. Pickering A was the first in a planned fleet of atomic power generators, teaming up with the four-reactor Bruce A station at Point Douglas on Lake Huron (where another four-reactor station, Bruce B, would be constructed in the 1980s). Ontario Hydro had the best talent

I

money could buy and lots of money to buy it. Foreign visitors curious about just how things got done at this world-class outfit were received at the brand new concave-glass headquarters right across from the seat of government at Queen's Park in Toronto. The office was just around the corner from the University of Toronto, where Kupcis had done his Ph.D.

If you were in an engineering discipline Hydro was, Kupcis said, a wonderful and exciting place to work. It "was known around the world for its problem-solving skills in operating systems." Kupcis, an unassuming man whose manner contrasted sharply with Hydro's brash image, had worked his way up to the nineteenth floor and the CEO's office in a firm dominated by engineers. He resigned in 1997 after commissioning a report into the problem-plagued nuclear division. Cracks had appeared in Pickering's rolled joints early on, and, as the report made all too clear, the whole enterprise was riddled with operational confusion. The most common metaphor describing his departure was somehow appropriate to an organization that had, from its beginnings, treated the natural world as a battleground. The quiet executive had, it was said, "fallen on his sword."

Soon after his departure the Ontario government was getting ready to sell off the most secure, moneyspinning part of what had once been Ontario Hydro. The corporation had recently been broken up into pieces, the better to prepare it for privatization and for the much ballyhooed opening up of the electricity market to commercialization. Plans for new nuclear generation had been scrapped years before. Kupcis was looking back at what had happened to the once-proud organization. "The passing of Ontario Hydro," he said, "was a non-issue. There were no big speeches. No one cared anymore."

🔊

On one level, the story of Ontario Hydro is simply a matter of an enterprise that was too big and too arrogant for too long. For Floyd Laughren, the provincial treasurer under the NDP government of the early 1990s, it was a case of the public company squandering its ample reserves of public support. In 1996, soon after a new government pledged to privatizing Ontario Hydro came into power, Laughren served on a legislative committee that was sent out around the province to look into what had become a nuclear fiasco. On the tour he noticed that both the people who presented their opinions to the committee and the people he met in the hotels and coffee shops were quietly angry with "that goddamn nineteenth floor." Ontario Hydro, he found, had precious few friends. The result by 2001 was that Hydro, after six years of preparation for commercialization, seemed set to disappear virtually without a peep. Laughren—first elected to the Ontario legislature even before Allan Kupcis joined Hydro—was a socialist supporter of pretty well everything public. He was astonished—much like Kupcis—that such a massive change had come to pass with so little opposition. But it would become apparent that while support for the institution that was Ontario Hydro had certainly faded over the years, anyone defending the principle of public ownership (not to mention regulation) of the province's electricity system would be tapping into a wellspring of popular sentiment.

Hydro's demise had everything to do with its genetic makeup. Its beginnings came at the turn of the twentieth century, and its rise in the decades thereafter. Its fall came in the *fin-de-siècle* period. In between its operations represented the central tenet of the industrial enterprise that is consumer capitalism. Hard-wired to its corporate DNA was a belief in the unending expansion of the energy economy: that more and more electrical energy would be required to fuel eternal growth; that nature could provide both limitless resources to do the job and a bottomless pit for the waste; and that all of this could be accomplished with the same splendid efficiency exhibited by an industrial food system that consumes ten calories of fossil fuel energy to generate a single calorie of food energy.

All of this raises the popular question of "sustainability." That is what another physicist, Joe Vise, was thinking in 1979 when he scaled a chain-link fence at Darlington, where Hydro intended to build its biggest ever nuclear generators. He and the other protestors, along with a growing body of thinking about energy planning, had warned Hydro that the megaproject constituted an expensive and dangerous folly. There was nowhere to put the toxic waste. It would be better, they argued, to invest the billions of dollars into snug buildings and wind power—to take what was being called the "soft energy path." Otherwise, all the recent talk about the ecological fate of the planet and the need for environmental protection would be just another contribution to the rising levels of hot air. What's more, the "energy crisis" that was concentrating so many minds in the wake of the Iranian revolution of early 1979 would remain with us in one form or another. Some twenty-five years later, the soft-energy-path types would be proven correct.

This book starts with that 1979 demonstration at Darlington. Joe Vise and his fellow protestors in the anti-nuclear movement were part of a mass upsurge that coincided with the demise of Keynesian economics and the rise of market fundamentalism. Ontario's Progressive Conservatives, a political party that by 1979 had been running the province for thirty-six years, had been using Hydro as a tool of their successful industrial strategy. During the premiership of William Davis (1971–85), his Big Blue Machine drove Canada's bold experiment with nuclear power. Davis and the members of his government were activists in their own way, bringing in Ontario's first real environmental laws. But Davis and company also ploughed ahead with Darlington, ignoring the warnings of the greens while slapping aside a handful of militant market fundamentalists within their own ranks.

The surprise defeat of the Tories in 1985 represented a historic political shift that offered the province the opportunity to chart a different course for Ontario Hydro and its grandiose expansion plans. But that new direction did not come about. The successive Liberal and NDP governments both claimed to be open to finding another way of planning the province's electrical future, but they were distracted by matters apparently more pressing. Meanwhile Hydro continued on the same roll that had long fuelled its growth—indeed, ever since the day its founder Adam Beck declared so famously, "Nothing is too big for us . . ." It embarked on an ambitious new plan to build more and more generators. It continued to assume an endless growth in the demand for electricity, hoping that a nod to snug buildings and windmills would appease the critics and give its political masters something to crow about. In a reflection that said much about the way the old Hydro operated, Kupcis called it "superb planning gone totally wrong" and "incredible arrogance."

When Hydro's grand plan for more generators was eventually abandoned, so too was the notion of any coherent planning for Ontario's electricity future—nuclear, green, or anywhere in between. Ironically, the nuclear generator expansion project was shelved on the socialist watch, when the New Democrats installed an ostensible environmentalist as Hydro's boss. Maurice Strong, a man fond of market-based solutions and determined to shake things up, drove a stake through the heart of Hydro's nuclear expansion division. He also stifled a conservation and renewable energy impulse that was just beginning to take root in the huge utility. He divided the firm up into units that could be separately sold off. Astonishingly, he asked Bay Street accountant Bill Farlinger to recommend a new path for Ontario Hydro. Farlinger was acting as the political godfather—and bagman—for the market-oriented, laissez-faire elements who had taken over the Progressive Conservatives. When the Tories under Mike Harris gained power in 1995, Farlinger succeeded Strong as chair of the Hydro board of directors. There followed a political train wreck the likes of which the province had not witnessed for generations. The route of destruction took several years to travel because the government underestimated the complex twists and turns of the journey, but Hydro was being legally uncoupled and readied for the private track just as the Enron saga unfolded south of the border. Meanwhile, Ontario's electricity system was also being readied for commercialization, for "market-opening." A new free market in power would surely lower the price of everyone's monthly bills—at least, that is what the province was told. Whether the privatization boosters ever believed it remains unclear.

By 2004 a new government was attempting to pick up billions of dollars of pieces scattered by the train wreck. The old Ontario Hydro had kept the lights on and the price down by shifting the debts to the future along with the environmental costs: dangerous air locally, global climate change, and nuclear waste who-knows-where. The Harris government's market-based vision had intended to do the same thing. Ontario was left to face the hard reality of summoning the political will to do things differently in the face of a public lured, as it had always been, by the siren song of steady power supplies at low prices.

For years Ontario's greens, shaped by wrestling with Hydro, had pointed out the folly of ignoring a soft energy path in favour of uranium, coal, and other non-renewable resources. But their warnings had little effect. That is because of another "hard reality." The champions of a soft energy path have been a fragmented lot. They have been "anti-nuke," to be sure; small is beautiful, definitely. But environmental activists have also focused variously on acid rain, clean air, pesticides, Kyoto ratification, and a dozen other important issues. They had to face attacks on environmental protection regulations by tax-fighting politicians who were not only boldly declaring this or that jurisdiction Open For Business but also determined to do away with "red tape." The Walkerton water contamination tragedy was one result of that predilection. Against this background, it is scarcely surprising that the environmentalists made so little progress against the champions of business as usual—the power-sucking corporations of the Association of Major Power Consumers of Ontario and the federal government's lavishly funded Atomic Energy of Canada Ltd. (AECL).

Ontario Hydro's demise, like its historical origins, was bound up, then, with a mixture of contending political and economic forces, leavened with more recent concerns for the environmental effects of how we produce and consume electricity. Those forces and concerns have preoccupied many critics of the company over the past two decades or more. One of them, Ralph Torrie, has followed the Ontario Hydro story from his days as an undergraduate physics student in the early 1970s, when he did co-op placements teaching reactor control-room operators at Hydro's Deep River training centre. Though he eschewed graduate studies for a career of writing energy-efficiency software and promoting the soft energy path, Torrie has maintained an understanding grounded in physical reality, not the abstract projections of energy-market economists.

Torrie has inevitably returned to what he calls the "bizarre notion" that the demand for fuel and electricity would continue to grow indefinitely; he calls for a physical explanation of how that could possibly happen. A fundamental law of classical physics, the law of conservation of energy (aka the first law of thermodynamics) states that the universe contains only so much energy—there is always the same total amount—and it can be neither created nor destroyed: it can just change form. It can be converted from heat to light to motion and back again, but never increased or decreased. This theory is what prompted Torrie, in 1976, to write *Nuclear Power and Future Society*, the first thorough critique of Ontario Hydro's manic power plans.

Almost thirty years later Torrie was still livid, this time about the proposal sprung in spring 2004 not only to spend more money attempting to revive Ontario's nuclear power plants but also spend more on building new ones. "There comes a point when you see the same mistakes that were made not that long ago about to be made again," he said. "You have to either laugh or cry." By that time the plans for a privatized, let-the-market-decide future for the province's power system had collapsed, and Ontario once again found itself at a fork in the road. It faced the breakdown of its electricity economy.

"This is our last chance to put Ontario on a path to a safe and environmentally sustainable energy future," Torrie said. "If we get it wrong this time, we will be in for a very difficult energy future in this province, environmentally and economically."

Instead of laughter or tears, Torrie chose the therapy of a letter to the editor to ease his frustration. His short screed declared that if all the money spent up to that point trying to re-tube the Pickering nuclear reactors had been given to him in five-dollar bills, he could have burned it to heat his house for a hundred years and still have a few million dollars left over, which he could then shred and use as attic insulation. Or, alternatively, the money "could have been loaned to Ontarians to upgrade the efficiency of their homes and appliances."

How well the province responds to this great challenge remains a matter that future readers of this book will be in a position to judge for themselves.

two

Power and the Tory Dynasty: The Making of a Crisis

A hard path can make the attainment of a soft path prohibitively difficult, both by starving its components into garbled and incoherent fragments and by changing social structures and values in a way that makes the innovations of a soft path more painful to envisage and to achieve.
— Amory Lovins, "Energy Strategy: The Road Not Taken?" 1976

AT A CERTAIN POINT on that bright June day in 1979 Joe Vise found himself staring at the foreboding fence that Ontario Hydro had erected around its latest nuclear construction site. He didn't look at all out of place. In fact, his beard and shock of curly hair made him fit right in. Back-to-the-landers and other members of the bib-overalls set had arrived from their Eastern Ontario acreages in an assortment of old sedans and vw vans. There were elderly Quakers in sensible shoes, babes in arms, and babes in tie-dyed slings. There were busloads of pacifists and anarchists from Toronto. One contingent had made the 80-kilometre journey from Peterborough by bicycle, camping overnight at the gathering spot, Darlington Provincial Park on the Lake Ontario shore. By noon the wardens were thinking of closing the park because it had reached its five-hundred-vehicle limit.

Despite the impressive size of the gathering, Vise knew as well as any of the others milling about that the crowd did not exactly represent the "ordinary Canadians" being then championed by NDP leader Ed Broadbent, MP for the nearby riding of Oshawa. Most ordinary Canadians tend to devote one of the first sunny weekends of the summer to the cottage. If they do get to a lakeside park, it is not to prepare for a demonstration. The people with the protest signs at

Darlington constituted the concerned and the committed—concerned about the spread of nuclear power, committed to stopping it.

Vise and the other demonstrators had been following the news over the past little while, and the news wasn't good. Hydro had just revealed that a heavy-water leak had forced the closure of the Nuclear Power Demonstration plant at Rolphton, northeast of Algonquin Park. There were also reports that Lake Ontario's herring gulls were exhibiting erratic behaviour, failing to protect their nests adjoining one of the world's most polluted freshwater lakes. More bad news: radioactive tritium had just been found in the water supply in Pickering, where Ontario Hydro already had a nuclear power plant. As a result there had been yet another spat in the legislature over Hydro's notorious deal—no tenders, no performance guarantees, full payment before delivery—to purchase boilers for the Pickering generators from Babcock & Wilcox. That was the same company responsible for designing the Three Mile Island plant at Harrisburg, Pennsylvania.

Just three months earlier, in March 1979, the nuclear station at Three Mile Island had come dangerously close to a catastrophic radiation leak. The word "meltdown" had suddenly made its way into the late-twentieth-century vocabulary. Now no one could avoid the obvious pun: the heavy political fallout from Three Mile Island resulted in the mobilization of people across Europe and North America. The demonstrators gathering at Darlington were part of an international weekend of civil disobedience targeting nuclear power plants.

Some eight thousand people had already assembled at Kalkar, West Germany. Like Ontario Hydro, the German utilities were counting on a dizzying expansion of nuclear power. The Germans hoped to have sixty reactors humming along by the turn of the century. The dour British magazine *The Economist* had opined that "this pre-Harrisburg calculation is now beginning to look silly."

Still, just because the more sophisticated elements of the business press were now less sanguine about nuclear power, that did not detract from the sense of urgency at Darlington. By day's end Joe Vise was among the people who had draped ragged scraps of carpet onto the barbed wire and scaled the fence to occupy the site of the proposed Darlington nuclear plant. He was the only nuclear physicist among them.

Vise had completed his Ph.D. dissertation in 1963 at Columbia University in New York, where he used the reactor at the Brookhaven National Laboratory to study an isotope that had a half-life of just less than one second. By 1979 he was teaching at the University of Toronto and running a low-energy particle accelerator in the sub-basement of the McLennan Physics Building. The work was part of a nuclear astrophysics research project modelling nuclear reactions in stars. For Vise the most pressing issue in Ontario Hydro's plans to dot various waterfronts around the province with multi-reactor nuclear stations was the apparently intractable problem of disposing of nuclear wastes that had a half-life of some 240 centuries: "What do you do with the garbage? The decommissioned nuclear reactors and the spent nuclear fuel?" Ontario Hydro was, supposedly, in the process of figuring that out. For one thing, Hydro was studying the Canadian Shield's stable granitic formations in search of a dump site. With his mind on sociology as well as physics, Vise countered, "The idea of maintaining secure structures for even a hundred years is unthinkable considering the instability of human governing systems."

In the late 1970s all that the public had to go on was the Canadian nuclear industry's disposal record, a sorry affair that the growing anti-nuclear movement viewed with anxiety. Just down the shore from Darlington was the town of Port Hope, where radioactive rubble and fill from Canada's main uranium refinery had been used as construction and landscape material. Far away in the North, where Ontario Hydro and its federal minders at the Atomic Energy Control Board wanted to deposit the poisonous nuclear waste, radioactive tailings from the mining operations that produced the nuclear fuel in the first place had simply been abandoned in open piles, to be washed away in wilderness watersheds.

All of these issues, together with the recent events at Three Mile Island, animated the people who had arrived at Darlington. What they saw there was a barren, 1,200-acre construction site surrounded by miles of chain-link fencing. The 500-kilovolt transmission tower just inside the fence was already festooned with massive banners—"Stop Darlington" and "Honk For No Nukes"—clearly visible from Highway 401. Peter Dundas, Paul McKay, and Peter Onstein had scaled the tower the day before in a stunt that succeeded in attracting media attention as a prelude to the planned occupation. The high-altitude protestors were equipped with a CB radio to communicate with the press—that is, until one of them dropped the batteries, forcing their backup man, Paul Gervan, to handle things from his van outside the fence.

The tower action had all the earmarks of a classic Greenpeace publicity affair, except that Greenpeace had nothing to do with it. Toronto's fledgling Greenpeace chapter had plans of its own for the day. Hydro security guards were as surprised as the other demonstrators when a Greenpeace-chartered Cherokee began disgorging a "Splat Team" of parachutists onto the site. With that, Joe Vise and the others who had decided to occupy the site began to clamber over the fence. One person was tackled before John Kearns, Hydro's head of security, intervened. Kearns realized that it was a containable situation. The main threat to the corporation was posed not by the protestors but by another blow to its image. The best way to avoid a lot of fuss and unwanted press attention was to wait until the demonstrators sat down before arresting them.

"They said they'd get forty people in," a Hydro official said. "I guess they made it." In fact, sixty-six people were arrested and carted off to jails in Oshawa and Bowmanville. Vise found himself facing the choice between signing an agreement to stay away from Darlington or spending a couple of nights in the lock-up. Believing that his fight was with Ontario Hydro and its nuclear plans, not with the police and courts, he signed and made his way back to Toronto. He had no intention of returning to Darlington anytime soon.

Much to his surprise, Vise did find himself back at Darlington a few years later. This time he went not as a protestor but as a tour-guide-cum-university-instructor. He was teaching a second-year undergraduate lab, and the course co-ordinator had arranged a field trip to the partially completed nuclear power station. The size of the construction bill almost matched the monumental nature of the place: Darlington's cost rose from an original estimate of $4.5 billion to over $14 billion when it finally began contributing full power to the grid in 1993. Vise and his students couldn't help but be amazed by the 71-metre cylindrical vacuum building that would be maintained at negative atmospheric pressure, the twelve-storey turbine hall that was over half a kilometre long, the four reactors whose 1.8-metre-thick reinforced concrete walls and heavily shielded calandria—the main reactor vessel—would contain 6,240 bundles of uranium fuel encased in zirconium alloy tubes. Particularly striking was the precision that could be achieved in the construction of something so immense. "The four generators sit on a different floor from the reactors," Vise said, "and you have these machines that are dynamically vibration-mounted, that can't tolerate vibration of more than a fraction of a millimetre in amplitude."

The designers of Canada's deuterium-uranium system and Ontario Hydro's engineers had succeeded in scaling up the country's first large research reactor, the NRX, a machine that emerged from the wartime nuclear push at Chalk River, Ontario, up the Ottawa River from Pembroke. Canada's nuclear effort had received a big boost from the U.S. military at the start of the Cold War. The story goes that Admiral Hyman Rickover, the father of the American nuclear navy, was frustrated when his underlings told him they had scoured the U.S.A. for a facility to conduct the fuel irradiation testing for the first nuclear submarine, the *Nautilus*. No luck. "There must be some place?" sputtered the Admiral.

"Well," said the project's senior metallurgist, "there's one little place way up in Northern Ontario where they have a reactor that can do this."

"Right! Let's buy it."

"We can't do that, sir. It's a national establishment."

"Well, then we'll just create a State Park around it."

The integration of the U.S. military's nuclear program with Canada's nuclear power efforts was so complete that Canadian scientists were not allowed to read the edited versions of their own research when it was finally published. The journals were classified. That did nothing to stop the government-funded research efforts—no business has benefited more from public largesse than the nuclear industry—from moving ahead. By 1962 Canada's Nuclear Power Demonstration (NPD) Plant at Rolphton, near Chalk River, was operating. The 20-megawatt NPD partnership between Ontario Hydro, Canadian General Electric (CGE) of Peterborough, and Atomic Energy of Canada Ltd. began in 1955. It was a high-stakes game that promised a win-win-win outcome for the powerful players. AECL, a Crown agency incorporated in 1952, could show off its research skills and the utility of nuclear research, justifying its expanding budgets. CGE wanted a foothold in an emerging industry that seemed likely to offer huge government contracts. Ontario Hydro wanted to generate more electricity, and more after that. Midget reactors like the ones designed for the *Nautilus* would not do; nor would the Rolphton test plant. Hydro was, according to one background paper, "in a hurry."

In the mid-1950s Ontario already had plans for a reactor with ten times the Rolphton output. That one would be built on the Lake Huron shore at Douglas Point, near the sleepy farm communities of Kincardine and Port Elgin. The giant utility had launched a frenzied "Live Better Electrically" advertising campaign urging Ontarians to be "completely modern" by using "flameless heat." It bought up enough land beside the lake to build two more plants, each with four reactors, and each dwarfing the original 250-megawatt Douglas Point unit. Then there would be another eight-reactor set-up at Pickering. As Toronto metastasized across the landscape that facility would soon become the world's only atomic power plant embedded in a metropolitan area. By the 1970s came plans for Darlington, with four mega-reactors of 900 megawatts each, the largest Hydro would ever build.

"It was fascinating, a *beautiful* piece of technology," enthused Joe Vise, recalling his second trip to Darlington. "You could see why any engineer working on it would be absolutely carried away." It was the bold mentality at Ontario Hydro and Ottawa's AECL that gave rise to the catchy name that the public companies would use for the reactors: Candu. Although it was technically an acronym for Canadian deuterium uranium, no one was in any doubt of the confident Can-Do message, and the companies went on to peddle the reactors to dictatorships around the world.

As a nuclear physicist Vise was well aware that no one knew what to do with the used irradiated zirconium alloy tubes. He had other reasons as well for climbing into the paddy wagon at Darlington in 1979. He knew that, in the same way that the Candu brand symbolized engineering hubris, Darlington represented everything that was wrong with the rapid expansion of technologies that centralize too much energy production in one place, rendering the whole electrical system more vulnerable than it needed to be. He also recognized the folly of basing electric power plans on a particularly crude assumption: that both the demand for and the supply of power had to keep rising.

"The whole approach that Hydro had started in the fifties and sixties was that you project gigantic demand in electric power," he said. "To make generation economical you keep promoting demand. You then have to fulfil the demand you helped create." For Vise and other critics, this was an irrational approach—and Darlington was its latest manifestation.

⏻

Even though it was owned by the government, Ontario Hydro was "a power unto itself." This observation was repeated so often that it became one of the great clichés of Ontario politics. Ever since the days of Adam Beck, the manufacturer and power broker who early in the century drove the establishment of the Hydro-Electric Power Commission of Ontario, the Crown corporation had become a behemoth, seemingly unaccountable to its government owners. But in truth Hydro, the business establishment, and the ruling Tories were bound together with close and strong ties. Between the 1950s and 1970s Robert Macaulay, once known as Mr. Energy, was variously Hydro's vice-chair, a Tory cabinet minister, and counsel for the Association of Major Power Consumers of Ontario (AMPCO), the industry lobby that ensured that the more power a company bought, the cheaper the power. Macaulay's father Leopold had been Tory leader prior to the beginning of the dynasty that governed Ontario from 1943 until 1985. His brother Hugh served as chair of Ontario Hydro after occupying the same post for the Progressive Conservative Party itself.

It was not, then, so much a matter of *not* being accountable. It was more a matter of *to whom* the utility was accountable—the legislature, the government, or the Government Party. Many of those who demonstrated at Darlington in 1979 had never known any government other than the party machine presided over by Premier Bill

Davis and his predecessors Robarts, Frost, Kennedy, and Drew. When George Drew took over in 1943, the provincial economy was hopped up on military spending that soon gave way to the storied postwar boom. The province's population more than doubled on the Tory watch. Per capita income tripled between 1941 and 1975. The two biggest areas of engineering and construction were highways and electric power facilities, both publicly financed.

Ontario laid the foundations of a welfare state, although the cautious government only very reluctantly signed on to medicare. Particularly under the Red Tory wing of the party represented by the people around Bill Davis, the government took an active role in managing the economy. None of the Red Tories would have disagreed with John Kenneth Galbraith, the most famous economist Canada has ever produced. Galbraith, who hailed from Iona Station in the Tory heartland south of London, remarked in his influential 1967 book *The New Industrial State*:

> No hungry man who is also sober can be persuaded to use his last dollar for anything but food. But a well-fed, well-clad, well-sheltered and otherwise well-tended person can be persuaded as between an electric razor and an electric toothbrush. Along with prices and costs, consumer demand becomes subject to management.

Galbraith adhered to the gospel according to Lord Keynes. During the Great Depression the influential British economist John Maynard Keynes argued that government spending could iron out the ups and downs of the business cycle by stimulating demand and boosting economic growth. The market was a necessary mechanism in the workshop of the economy, but not enough in itself. For the Ontario government no instrument of state was as instrumental at stimulating demand as Ontario Hydro. With the demand for electric razors and pretty well anything else electric seemingly growing without pause, the company's appetite for expansion matched that of the provincial economy. The good times rolled by as quickly as the water heaters, dryers, and space heaters rolled off the assembly lines scattered across the southern part of the province. People embraced modernity, living better electrically in their new all-electric homes. Ontario Hydro abandoned its wartime conservation campaign and, together with its allies in the appliance industry, began pushing consumers to go out and purchase the new plug-in gadgets.

Postwar government spending began with 1950s megaprojects, including a second hydro generating station at Niagara (Beck II) and a massive new dam that was part of the St. Lawrence Seaway development. Hydro also launched an ambitious expansion of coal-fired generators whose stacks belched acid-laced fumes across a landscape increasingly cluttered with high-voltage transmission lines. For the first time Ontario Hydro was coming to rely on non-renewable fuel sources to generate power. Fossil-fuel plants with a capacity of 12,000 megawatts were approved. The largest of them, on the Lake Erie shore at Nanticoke, was a coal-fired installation that could generate almost as much power as the entire eight-pack reactor set at the Pickering nuclear complex. But for the managers and engineers whose build-it-and-they-will-come philosophy drove Hydro expansion, the future would always lie in nuclear power.

Canada's wartime "Minister of Everything," C.D. Howe, created AECL in peacetime as a research and development organization, and organized the corporation in a way that would ensure maximum cooperation between the designer (AECL), manufacturer (Canadian General Electric), and customer (Ontario Hydro). It was, said historian Robert Bothwell, "a clear industrial strategy of a kind often sought but not usually found in Canada." In 1958 AECL got a boost from the federal Diefenbaker administration, which was desperate to be seen supporting Canadian industrial and research leadership after the public uproar over the cancellation of the Avro Arrow jet aircraft project.

The Candu reactor was the proud product of a whole generation of scientists who saw themselves as leaders of a northern nation's nationalist project. "We've suffered enormously in this country from being hewers of wood and drawers of water," said AECL chair Ross Campbell just as opposition to nuclear power was starting to build. "And now we've got an answer. Suddenly this answer is being knocked. I personally think it's a tragedy."

In 1962 Robert Macaulay managed to catch the spirit of the whole enterprise when he assured the leaders of the municipally owned utilities responsible for local power distribution that there was no limit to the promise of nuclear power. Now, with the Rolphton demonstrator up and running, he said, "The day will come in our lifetime when electricity will be generated directly from an atomic unit the size of a water glass." Perhaps Macaulay made the mistake of believing his own propaganda. Maybe he was simply deluded. Whatever the case, nuclear power meant larger centralized generating stations. And it was nuclear power, coupled with the end of Keynesianism, that would eventually lead to the decline and fall of Ontario Hydro's electric empire.

It was a story bound up with the shifting fortunes of the Ontario Conservatives, who had the good luck to govern the province during a period of unparalleled prosperity that lasted until the mid-1970s. The remarkable dynasty was based in part on the Tory image of being prudent managers, which was not a hard reputation to cultivate in good times. The Tories also owed their longevity to their knack of changing leaders before the incumbent became tired and stale. Another factor was recruitment. Any successful institutional machine has the ability to cultivate knowledgeable and skilled up-and-comers. When Leslie Frost, the man they called "Old Man Ontario," retired in 1961, he was succeeded by corporate lawyer John Robarts, a self-described "management man." Macaulay, who had also run for the top job, tossed his convention support behind Robarts and got his reward with an appointment as minister of commerce and development—which was in addition to the posts he already held as minister of energy resources and first vice-chair of the Hydro-Electric Power Commission of Ontario.

Macaulay in turn found a job for the young lawyer who had managed his leadership campaign. Bill Davis, first elected in 1959, could not afford to make politics a full-time career, and Macaulay had him appointed as second vice-president of Hydro, a job that more than doubled his salary as a member of the provincial legislature. One day Davis and his mentor stood in Macaulay's Queen's Park office, gazing south at the Hydro building on University Avenue. Macaulay said, "Those guys think they run the government!"

Whether the tail was wagging the dog or vice-versa would provide fodder for academic theses and legislative inquiries. What was clear was that the relationship, though not always smooth, was symbiotic. What government would not take advantage of a Crown corporation with such massive financial clout? How could a company so big and so powerful not have an influence on affairs of state and the direction the province was moving?

After Bill Davis became premier in 1971, things appeared to be unfolding as they always had. Within a year Davis was proudly announcing the commissioning of the first nuclear unit at Pickering, an event likened to the birth of "a bouncing new baby." But signs of trouble were already on the horizon. Pickering A had been approved even before the much smaller prototype at Douglas Point was finished. Hydro had originally planned the operation as two nuclear reactors, but soon doubled the facility's capacity to four 500-megawatt units that would cost $528 million in total. When the final switch was thrown in Pickering A in 1974 the cost was $746 million—$218 million over budget. Few suspected that these numbers—already massive by the standards of the early 1970s—would be dwarfed by the billions that Hydro would eventually pour into the black hole that was Pickering A.

By the time Pickering opened, the world of energy had been knocked off its axis. In 1973 the Organization of Petroleum Exporting Countries (OPEC) unilaterally announced a fourfold increase in the world price of petroleum. Panic set in. The International Energy Agency predicted that oil would cost $100 a barrel by the turn of the century. Industrialized countries reacted to this "oil crisis" in two apparently contradictory ways. Echoing the frugality of wartime rationing, governments began talking about conservation. They lowered highway speed limits and urged people to switch something, anything, off. Inflation rose along with interest rates. The economy went into recession.

A common theme in the blizzard of policy papers and learned analyses that emerged from the Oil Shock was that nuclear power would have pride of place in supplying electricity to the energy-starved nations of the industrialized world. The years 1973 and 1974 saw a huge spike in orders for nuclear power plants in the United States, some forty each year. After all, hadn't U.S. Atomic Energy Commissioner Lewis Strauss announce in 1954 that nuclear electricity would be "too cheap to meter"? The four short words would live on to haunt the nuclear industry, just as Ontario Hydro's commitment to "power at cost" would keep coming back at the company.

The sudden nuclear imperative was a boon to Hydro's heady expansion plans. In May 1973—months before OPEC's oil price hike— the company's annual report predicted that nuclear power's contribution to provincial electricity production, then at 9 per cent, would hit 60–70 per cent by 1990. In the following month Premier Davis announced approval of a $3.5-billion Hydro expansion plan and changed the name of the corporation to Ontario Hydro. In the wake of the Oil Shock a Hydro planning study indicated that the utility was basing its plans on a 7 per cent annual growth in electricity demand and was intending to meet that growth exclusively with uranium and coal. Despite its own role in actively encouraging power demand through twenty years and millions of dollars of "Live Better Electrically" campaigns, Hydro now maintained that demand was something "over which we have no appreciable degree of control."

The Hydro name change was accompanied by other legislative measures. Among them were changes to the law governing the Ontario Energy Board (OEB), a regulatory body that did not regulate Hydro but had the power to scrutinize rates. The government established a new Ministry of Energy, which did not have the power to arbitrate any conflicts between Hydro and the OEB even though the minister was ostensibly responsible for both. In 1973, around the time Davis announced the massive Hydro expansion, a frustrated Donald C. MacDonald, the first Ontario NDP leader and a veteran parliamentarian, asked the new energy minister, Darcy McKeough, a series of hypothetical questions. Suppose someone who believed that Hydro's plans were too ambitious and too accelerated took their argument to the OEB? Suppose the Board became convinced that this was indeed the case? Suppose the minister were to recommend a change in plans to Hydro?

"Hydro can, in effect, do as it pleases? . . ." MacDonald asked, perhaps rhetorically.

The answer from McKeough was to the point. "I think that what the member . . . has pointed out is quite correct."

This lack of regulation did not mean a lack of public scrutiny. On the contrary: in the early 1970s the utility began to face fundamental questions raised by a press that, high on the Watergate scandal south of the border, was on the lookout for scandal. "Hydro-gate" broke in 1973 when it was revealed that a Davis friend, developer Gerhard Moog, had received a sweetheart deal in connection with the construction of the giant new headquarters that would house Hydro's burgeoning bureaucracy. The opposition in the legislature raised a clamour over the cost and magnitude of Hydro's plans. More trouble came in 1975, when Hydro proposed a rate increase of nearly 30 per cent for the following year. By that time the premier had already made a move that was time-honoured in Canadian politics. He appointed Arthur Porter to lead a royal commission that would investigate electric power planning in the province, probing Hydro's demand forecasts and nuclear program. It was to be the first of many outside inquiries, but the attempt to buy time did not prevent the Tories from slipping into minority government status in the election of 1975.

Meanwhile, Hydro continued to stalk the province like a man determined to use a sledgehammer to pound in tacks. The utility was looking for sites for new megaprojects, even though the old certainties about growing demand were quickly evaporating as inflation and interest rates crept upward along with public concern about matters environmental.

The nuclear establishment, founded on a faith in science and sustained by a national commitment to research and development, was unprepared to meet the changing climate of opinion. When AECL built the model town of Deep River in 1944 as a place to house its Chalk River employees, the company brought together over one thousand scientific workers and set them down in the wilderness, far from spying eyes. Years later, buoyed by a postwar popular enthusiasm for scientific marvels, the Crown agency was institutionally ill equipped to deal with the sudden suspicion of science, and especially all things nuclear. While the Best and the Brightest working at the Chalk River lab looked inward, secure in their own rectitude and proud of their international reputation, the world outside was asking questions about nuclear waste and watching *The China Syndrome* on the silver screen while Three Mile Island dominated the headlines.

Even before Ontario Hydro built its fence around the Darlington construction project, the self-described "Darlington Dozen" had invaded the site. They waded through the mud not long after the company broke ground for its newest nuclear plant. Paul Gervan and Greenpeace's Dan McDermott (who would be a member of the sky-diving Splat Team two years later) were among those photographed planting soggy seedlings that rainy fall day in 1977. Hydro still had a lot to learn about image management. News photos taken that day showed a security man in a yellow outfit stomping the seedlings into the ground. The words "Ontario Hydro" were clearly displayed all over his hard hat.

"We ran the gauntlet of construction workers threatening to bust our heads," said Gervan, who was working at the Kingston Energy Conservation Centre at the time. "All we were doing was planting trees."

Within a year activists had returned to nearby Darlington Provincial Park. They had organized Sun Day, an event designed to help place renewable energy technologies on the map. A thousand people showed up to listen to folk music, sample hot chocolate prepared using a solar heater, and listen to speeches from members of the Ontario Coalition for Nuclear Responsibility, Non-Nuclear Network (aka Birchbark Alliance), and "CANTDU." Among the people sitting on the grass was Ralph Torrie, a young physicist who had prepared a 450-page submission to the Porter Commission dissecting the nuclear industry's claims that it was paving the way to a safe, clean future. Like many others, Torrie was worried. His concerns ranged from technical issues to the assumptions of the consumer culture. He considered how Ontario Hydro was betting the future on huge nuclear generating stations.

> Is this wise planning? Is it planning at all? Is high quality electricity the answer to our energy needs? What types of social change will be brought about by the all-electric society? Who will benefit and who will pay? What are the long term implications to our survival of developing high energy technologies which encourage continued exponential growth of nonrenewable resource consump-

tion? Is the nuclear electric future possible? Can we afford it? Do we want it?

To suggest that there might possibly be another path, that growth as an end in itself was the ideology of the cancer cell, amounted to heresy. But by the early 1970s there were already a lot of heretics about. They were not terribly enamoured of a prosperity defined by endless increases in supply and demand. They saw themselves as biological creatures, enmeshed in a web of ecological relationships—which included, sadly, fish swimming in acidified lakes and herring gulls starting to behave in curious ways.

The environmental movement was sophisticated enough to realize that successful politics demanded more than opposition. You had to have a vision, an alternative, something hopeful to offer. The struggle was an uphill fight. Not only were the critics ranged against Ontario Hydro and Canada's nuclear establishment, but they were also facing the flash and grab of consumer culture. Capitalism was innately connected to the imperative of endless expansion. Many greens were arguing, seemingly endlessly, that the single-minded pursuit of more power was a recipe for environmental disaster. A more benign approach, they said, would moderate demand and invest in renewable resources. They were fond of quoting Robert Frost:

> Two roads diverged in a wood, and I—
> I took the one less travelled by,
> And that has made all the difference.

The road less travelled, they said, involved a different energy strategy, a "soft path" that would lead to a "conserver society." The consumer society was following a "hard path," the Freudian figure of speech used by soft path advocates critical of the macho engineering lingo of taming rivers and harnessing the atom. The soft path was not simply being promoted by starry-eyed dreamers discussing wind power at various Sun Days. Ralph Torrie's detailed 1977 research study gave Ontario Hydro executives fits because it confronted them on their own terms with its technical analysis of the first law of thermodynamics and the costs and risks of the nuclear fuel cycle.

Not all of the critics were outsiders. Torrie's introduction quoted the 1976 resignation letter of a General Electric nuclear engineer who abruptly left a thirty-three-year career with the statement, "Nuclear power has become a 'technological monster' and it is not clear who, if anyone, is in control." In 1977 the Science Council of Canada, an official advisory body to the federal government on matters scientific, laid out a detailed vision bearing the unwieldy but suggestive title "Canada as a Conserver Society: Resource Uncertainties and the Need for New Technologies." A guiding light of the critics was physicist Ursula Franklin, professor of metallurgy at the University of Toronto. In the early 1960s, when the military was still detonating nuclear bombs in the atmosphere, Franklin went around collecting baby teeth to check for strontium-90.

The Science Council was prescient. It recognized the promise of smart boxes that used virtually no electricity—devices that would fine-tune the workings of much larger machines, saving energy and resources. The Science Council emphasized the importance of building energy savings into the design of better-insulated buildings. Given that Canada's suburbs had become places where the purchase of a loaf of bread required a short trip using an internal combustion engine, urban design was crucial. So too was a shift to decentralized and diverse forms of generating electricity. Small-scale solar and wind technologies use readily available and free renewable resources. Avoid hub-and-spoke systems, the critique said. Instead, place sources of power closer to where that power was going to be used; doing so would reduce transmission losses, for the further you send it along the wires the less useful is the energy when it reaches its destination, and the more it leaks away in the form of unwanted heat and electromagnetic fields.

Wind power and solar power would be more efficient and less environmentally harmful than power developed using coal and uranium. They would also allow for the possibility of public participation. A conserver society based on a soft energy path could be more easily influenced by citizens and citizen groups. In contrast to the Mackenzie Valley pipeline, the tar sands, and other remote megaprojects on the drawing boards in the 1970s, diverse energy systems could be tailored to local needs. Instead of extravagant, complex operations like the Darlington station, point-of-use energy conservation would allow and even encourage local involvement. Power-users would find it easy to imagine participating in a group planning the development of a few nearby windmills; they would find the idea of participating in planning for a nuclear power plant laughable. Indeed, Hydro went to the trouble of bulldozing massive amounts of earth to construct a huge berm both to conceal the Darlington station from passersby on the 401 and to protect the reactors from a main railway line that carried dangerous explosive chemicals every day. The Science Council pointed out:

> If the system is remotely and impersonally managed, or appears so to the user, the user will feel no share of responsibility in its operations. Any difficulties, any shortage of supply, will be shrugged off onto "the system," never to the fact that the user and his/her neighbours have . . . all been switching on their stoves during peak time. Whatever the system provides will tend to be wasted by individuals, who see no obvious connection between what they use, and the costs that fall on others or indirectly on themselves through taxes or the environment.

By the late 1970s this kind of thinking, stimulated by a second OPEC price hike, was making its way into public consciousness. There was a generalized sense that resources were limited. The sudden success of *Harrowsmith* testified to the tenor of the times: the magazine, bursting with wood-stove ads, boosted hands-on energy saving projects and organic living. The conserver society ethic even filtered up into official circles. Prince Edward Island declined to participate in the establishment of a nuclear power plant in New Brunswick, even though federal cash made the deal attractive. P.E.I. premier Alex Campbell reacted angrily to the suggestion that putting windmills atop schools in Summerside amounted to a "piddly little thing." "The energy establishment in this country could do with a good bit of shaking up," Campbell said. "As premier, my mission now is to urge Canadians to start taking a serious look at 'piddly little things.'"

In Ontario the government passed an Environmental Assessment Act intended to anticipate the potentially harmful effects of major projects, but immediately exempted the Darlington project. Ottawa established an activist Office of Energy Conservation that projected zero growth in energy consumption by 1990. The feds seemed to be taking the soft path more seriously, offering direct grants to home-owners and funding local energy conservation centres that often became hotbeds of anti-nuclear activism. Ottawa's official energy strategy target even foresaw a *reduction* in year-to-year energy growth. Prime Minister Trudeau asked Amory Lovins to lunch.

Like Joe Vise, Ralph Torrie, and Ursula Franklin, Lovins was trained as a physicist. The energetic American had a way with words— he coined the phrase that using electricity to heat your house (as Ontario Hydro so often recommended) was like using a chainsaw to cut butter. Some three years before Three Mile Island he was point-ing out, "Many in the $100-billion quasi-civilian nuclear industry agree that it could be politically destroyed if a major accident occurred soon." Relying on coal to generate electricity, he argued, meant that atmospheric carbon dioxide would lead to "perhaps irreversible changes in global climate" by the early twenty-first century.

Even before these warnings appeared in an influential 1976 article in *Foreign Affairs*, Arthur Cordell of the Science Council had brought Lovins to Ottawa. "He was our kind of guy," recalled Cordell, who at the time was preparing cabinet documents as part of his work on the conserver society file. "He used to walk around with a calcula-tor on his belt." Cordell knew that the push for innovative policies needed champions who in turn needed legitimizing references. *Foreign Affairs* fit the bill, and Lovins agreed with the Office of Energy Conservation that a modest or even zero growth rate in energy use was a realistic long-term goal for countries that opted not for big, brittle technologies within a narrow range of possibilities but for modest, thrifty approaches across the board.

For soft path advocates, "low" technologies were not unsophisti-cated, but they were simpler to understand and use. Stricter building codes reduced energy use. Structures could be designed to stay warmer in winter, cooler in summer. Wind turbines would be rela-tively cheap and low-risk, relying on smaller, simpler components that would be easier to standardize and manufacture. Roof insulation could compete with Arctic gas, passive solar energy could overcome nuclear power. But for all the differences between the two paths, Lovins mused that the most profound could well be in their sociopolitical impact. The hard path favoured not just centralized generation, but centralized authority.

> In an electrical world, your lifeline comes not from an understandable neighborhood technology run by people you know who are at your own social level, but rather from an alien, remote, and perhaps humiliatingly uncontrollable technology run by a faraway, bureaucratized, technical elite who have probably never heard of you. Decisions about who shall have how much energy at what price also become centralized—a politically dangerous trend because it divides those who use energy from those who supply and regulate it. The scale and complexity of electrical grids not only make them politically inaccessible to the poor and weak, but also increase the likelihood and size of malfunctions, mistakes and deliberate disruptions. A small fault or a few discontented people become able to turn off a country.

Those who champion the hard path, wrote Lovins, "peer myopically forward, one power station at a time, extrapolating trend into destiny by self-fulfilling prophecy with no end clearly in sight."

The same sharp contrasts of the future were being drawn at Ontario's Porter Commission into how the province was planning its electrical future. But Ontario's soft path boosters still faced the hard reality that had to do with concentrated power of a different sort. As the Science Council lamented, "Unfortunately, the future has little economic or political power."

📾

After Ontario's Conservatives lost their majority in 1975, and just as the Porter Commission was starting its work, Premier Davis was forced to appoint a select committee of the legislature to probe Hydro affairs. The result was that the swashbuckling utility suddenly faced an unprecedented level of public scrutiny. Critics lined up to tell the legislators that Ontario Hydro should take the soft path.

The select committee included rookie Liberal David Peterson and was chaired by the cheerful Donald MacDonald, who recalled the Hydro hearings as the most satisfying experience in his decades at Queen's Park. MacDonald, never known for being a strident partisan, would later recall that in giving a speech, whether in Kapuskasing or Texas, his old rival Bill Davis was sure to include a favourable mention of two things: his home town of Brampton and nuclear power. MacDonald saw Davis not simply as an ardent advocate of nuclear power but as "a crusader." Progressive Conservative stalwart Darcy McKeough, for instance, did not remember the decision to authorize the construction of Darlington ever coming before cabinet; the former treasurer could only assume that the premier had made the decision on his own.

Despite the premier's enthusiasm for a nuclear future, there was no way his government could avoid the gathering storm around Hydro. The Tory reputation for prudence was colliding with Hydro's expansion plans, to say nothing of eye-popping rate increases that would total 80 per cent between 1976 and 1980. The rate hikes stood in sharp contrast to the "Live Better Electrically" years. Between 1954 and 1967 there had been no general rate increase. In a twenty-year building plan issued just before the government was reduced to a minority, Hydro estimated that it would need $90 billion to meet its demand forecasts. This prompted McKeough, as the minister responsible for provincial finances, to do something that no provincial treasurer had ever done. He warned Hydro in 1976 that if it did not curb its growth forecasts he would deny the government guarantees crucial for the bond issues that the utility needed to float.

Within months the select committee recommended that Hydro institute a full conservation program. The acronym DSM—demand-side management—started to creep into the vocabulary of Hydro's critics. The idea of DSM, in this case applied only to the electricity sector, was simple: instead of building more nuclear and coal plants, why not concentrate on the demand side of the meter by building "conservation" power plants?

The DSM approach would not be taken up seriously by Ontario Hydro for another ten years. Meanwhile, controversy over Darlington mounted. In response to the government's go-slow directives, Hydro began to delay its plans, but between 1976 and 1980 still borrowed $7 billion to fund expansion. In 1977 there was no growth in the consumption of electricity as power-users responded to higher prices and the general sense that an energy crisis was hurting industrial economies. In that same year President Jimmy Carter described the challenge of changing U.S. energy policies as "the moral equivalent of war," or MEOW. By year's end Hydro was finally edging its demand projections downward, but not nearly as much as the select committee and energy critics were urging it to do. Hydro still hoped to continue building as if demand would keep growing at 7 per cent, but the Porter Commission's interim report of 1978 predicted a 4 per cent growth rate, urging Hydro to be more flexible and not to rely on large-scale nuclear stations to meet its demand forecasts.

The Porter Commission's interim report marked the first official, non-partisan warning about the soundness of Hydro's planning assumptions. Hydro again lowered its growth forecasts, and within a year it had begun to cancel plans for new plants and to mothball coal and oil burners already under construction. In 1979, as the anti-nuclear clamour rose with the accident at Three Mile Island, Ontario's Energy Ministry issued a turn-of-the-century demand forecast almost identical to a prediction that the environmental watchdog group Energy Probe had come up with three years previously. Both forecasts were 12,000 megawatts below the amount cited in Hydro's plans—or the equivalent of three four-reactor Darlington power stations. In 1980 the Porter Commission concluded its exhaustive five-year study of Ontario electricity with a final report calling for the province to reject Hydro's plans to extend its chain of large-scale plants. The Commission recommended that "the rigidity of supply planning, with its fixation on large-scale nuclear plants, be abandoned for the flexibility of demand management and smaller-scale additions to generation capacity."

The critics who appeared before the Porter Commission pointed out that making power cheaper for the larger users of electricity (such as the industry lobby AMPCO) encouraged waste and made those who practised conservation feel like dupes because others were paying less for using more. They also argued that those who used electricity, whether for nickel-smelting or desk lamps, were not interested in buying power—that power is a means to an end—heat and light. Soft path advocates urged energy planners to begin their work by considering the end uses of power and how those end uses could be served by using less power. It was a mistake to assume that nature could ever support anything close to an economy based on the ever-expanding growth in consumption of non-renewable resources. It was not, to use a word that would soon become fashionable, sustainable.

Still, all of these arguments did not seem to matter very much in Ontario. Hydro and its government remained committed to a scenario of endless growth in electricity consumption. The utility did finally lower its demand estimates in 1981. Its peak demand forecast of 38,000 megawatts by the turn of the century represented a whopping 52,000-megawatt reduction in the (mostly nuclear) expansion that Hydro had deemed necessary only three years previously. It was, however, 50 per cent above what consumption would turn out to be.

No nuclear plants were ordered in the United States after 1978, and most of the orders placed for plants after 1974 were subsequently shelved. The cancellations even included plants up to 97 per cent finished, abandoned by companies and appropriately named "IOUS." These private-sector Investor Owned Utilities began to sink into a sea of debt as the immense costs and long lead times of nuclear power became more obvious and opposition to nuclear power more concerted. There was no more talk about nuclear power being too cheap to meter.

In the United States—and events south of the border have a way of reverberating in Canada—President Carter's moral-equivalent-of-war speech gave rise in 1978 to a series of national energy policy acts, including the crucial Public Utilities Regulatory Policies Act. The groundbreaking PURPA empowered regulators to force electricity monopolies (both private or, like Ontario Hydro, public) to purchase power from independent companies. This caused much gnashing of teeth among the monopolies because it made them buy power from little outfits they did not own at prices set by public regulators they could not control. PURPA opened the door for independent operators who could "co-generate" electricity and heat, produce electricity from wind or other renewable sources, or simply generate a small amount of electricity for their own purposes. These NUGS, or non-utility generators, could sell the extra power they produced back to the main electricity grid. This opened the doors to small-scale technologies, the "piddly little things" like windmills atop schools, long scorned by Ontario Hydro and the other electricity heavyweights.

This stormy period was characterized by stratospheric interest rates, double-digit inflation, mass unemployment, and depression. It was also marked by the elections of Margaret Thatcher and Ronald Reagan, heralding the end of the Keynesian era and a resurgence of freebooting, unregulated capitalism. Ottawa's National Energy Program failed in the face of the 1982 oil price collapse and a storm of political opposition from corporate interests whose clout made the protests of anti-nuke activists look truly piddly. In the years that followed, mention of the NEP would be met with derisive references to the "real world" from mainstream pundits and other apologists of laissez-faire. After a generation of state intervention, the market would once again assume pride of place as private business recaptured the political ground it had lost in the decades after the Depression of the 1930s.

Ontario Hydro, long an instrument of interventionist government policy, was not prepared for the economic and ideological turbulence that hit in the 1980s; nor was the government. The high interest rates and recession forced a delay in construction at Darlington. Although Hydro still had nuclear ambitions, the corporation was awash in surplus capacity and planned no new generators after Darlington. Hydro still had so much power it had to try to export it to the United States. In 1981 the Conservatives were trying for the third time to regain the majority that would secure their place as the longest entrenched government outside of the Communist Party of the Soviet Union and Mexico's Institutional Revolutionary Party. But the Ontario Tories did not embrace the revanchist creed of Thatcher and Reagan. Rather, the run-up to the 1981 election seemed to echo Lenin's dictum that Soviet power plus electrification equalled Communism. Bill Davis and company ran on something called BILD, a grab bag of highway and other projects that included electric "energy centres." Ontario, it seemed, would keep on building big power stations. This new Board of Industrial Leadership and Development would, the government promised, "cultivate the end-use market for electricity . . . and facilitate the long run expansion of the indigenous nuclear-based electricity system."

Even if it was like cutting butter with a chainsaw, the government promised to once again promote residential electric heating. If Hydro still had a power surplus, it would export it to the United States, all the while importing the coal that helped produce the power glut in the first place. Davis also announced the acceleration of construction at Darlington. Selling the premier as someone you could trust in uncertain times, the party won the election with ease, regaining a majority.

Yet the premier himself had cut his political teeth in the salad days of a postwar boom that was now past. "Davis had no fresh vision to offer," wrote veteran journalist Rosemary Speirs. "He ran on the strength of his personal popularity—the known quantity, bland, unexciting, solid, someone people could trust to mind the store."

Joe Vise, like many of the protestors who gave the anti-nuclear movement of the 1970s its mass character, moved on to other projects. He immersed himself in anti-apartheid activity and nodded approvingly later on when his daughter was arrested during protests over the testing of the U.S. Cruise missile in Canada. He was often seen pedalling an old bicycle to the University of Toronto even in the coldest, sloppiest weather. When it came to heating the old frame house he and his wife Mary purchased on a hardscrabble farm northeast of the city, Vise was committed to the chainsaw and the splitting axe. He spent countless days in the tiring but rewarding job of cutting and stacking. He didn't have to read small-is-beautiful articles in *Harrowsmith* to understand the old cliché about the efficiency of heating with wood. It warms you up twice.

three

The Electric City and Public Power: Promise, Peril, and Peterson

Across rivers and chasms, over mountains, through forests, out into deserts—the slim wires run—bringing light, power and life. Into every corner of the country, electricity has sent its probing, creative fingers—developing farms, mines, industry and cities.

Electricity has been . . . and will continue to be . . . man's indispensable aid in breaking through the physical frontiers of the world. But now, with the full advent of the science of electronics—electricity begins to open the frontiers of the mind.

— from a Babcock & Wilcox magazine ad, 1950s

WHEN THE LIGHTS went out across Southern Ontario and most of the rest of the northeastern corner of the continent in 2003 and left millions without air conditioning on a scorching August afternoon, Ontario's premier suddenly began talking about conservation. Ernie Eves declared that he had now started to run the dishwasher only when it was full.

"Sometimes wake-up calls like this help everybody," the premier explained.

Another cliché immediately became commonplace: "Electricity is something we take for granted."

Except that it took a few days for people to hear those words from their usual sources of information. Electricity had not just made air conditioning possible. It powered radio, movies and television, microphones, amplifiers and loudspeakers, central processing units, monitors and modems. Electricity had energized mass culture and mass experience to such an extent that it had become like the air, something you only notice when it is no longer there.

By the turn of the twenty-first century it seemed perfectly natural to witness the bombing of Baghdad in real time, to follow, albeit fleetingly, the immediate travails of its power-deprived citizens sweltering in forty-degree heat. We have trouble imagining a world in which the streets remain dark at night, where there are no overnight celebrities. Electricity had clearly become more than electrons moving at lightning speed through a complex network of wires. It had helped to transform how people live and work and feel and believe. Electricity was vital to what Henry Wallace called "the century of the common man."

Back at the turn of the twentieth century electricity was only starting to make its way into affluent homes. Electric trolley cars appeared on main streets to replace horse-drawn transit. Cities competed to be the first to have those streets lit with clean, bright light. In the small Ontario city of Peterborough, the town fathers came up with the idea of "the Electric City," claiming that theirs was the first in the Dominion to boast electric street lighting. During a period when Montreal and Toronto were still using a toxic mixture of carbon monoxide and hydrogen called coal gas to light their streets with smoky yellow flames, Peterborough's downtown George Street was illuminated by seventeen electric arc lights.

The turning on of those lights, on Victoria Day evening, 1884, was a proud first for the city. When the lights on George Street went on, electric lighting was still a spectacular highlight at exhibitions and fairs in Europe and North America. People gazed in awe at the bright lights of big-city progress. During a period when the innovation was restricted to the wealthy, it was usually only opera houses and high-tone restaurants that boasted electric light. Indeed, when Thomas Edison, the quintessential American hero of the Gilded Age, inaugurated his first central-station electrical system in 1882, he placed his generator on Wall Street. As one writer noted, "It was not by coincidence that [Edison] focussed upon offices owned by men such as J.P. Morgan, who had the funds to finance Edison's project—and did."

Peterborough was a pioneer of Canada's industrial revolution because its fast-flowing Otonabee River offered ample supplies of the novelty that was hydroelectric power. When the potential of power at Niagara was just starting to spark a controversy over who would control the storied Falls, the Peterborough Light and Power Company had already installed a generator on the Otonabee's Dickson "raceway." The company secured the municipal street lighting franchise and lit up George Street. Peterborough's surplus power meant that by 1907 it was generating more manufactured goods than any city in Ontario on a per worker basis. Cereal from Quaker Oats (founded 1900) and canoes from what would become Peterborough Canoe became national brands.

Peterborough was even further ahead of its time when it came to luring new businesses. At the end of the twentieth century cities and provinces were falling all over themselves with offers of low taxes and cheap land to lure this or that employer to set up shop. Well before the end of the nineteenth century, Peterborough had accomplished just that. In 1890 the city fathers told the Edison General Electric Company that the town was in a position to come across with land valued at $18,372, municipal services worth $12,138, and a ten-year tax holiday. The U.S. company had initially built its works in Sherbrooke, Quebec, but immediately decamped when it got an offer it couldn't refuse. The men of property who constituted Peterborough's ratepayers were similarly enthusiastic. They authorized the deal by a vote of 656 to 11.

Peterborough had more than cheap land. The city sat on the Otonabee River at the foot of a series of fast-moving channels. The millraces on the river had already made the town an ideal location for the water-powered sawmills that processed timber from the dense forests of the Kawartha Lakes district. When Edison arrived, the company received fresh tax incentives to build two new powerhouses on the river. The company that would become Canadian General Electric emerged as one of the country's leading manufacturers of electrical apparatus. Well over a hundred years later its huge Peterborough plant remains a large industrial employer, though it has downsized in recent decades. It once produced everything from locomotives to refrigerators, from insulated wire and cable to huge hydroelectric generators. It also emerged as a key supplier to the nuclear industry, making refuelling machines and the fuel bundles that form the guts of nuclear power plants. Today its products are more limited (though the nuclear section with its fuel bundles remains a mainstay), but in spring 2004 the plant was completing the first of six generators—each the size of a railway car—to be used in portable power stations in Iraq.

🔱

Electrification was, from the first, a costly undertaking, and the earliest electricity systems were developed by powerhouses of the financial variety. Still, just as hydroelectric power transformed Peterborough into a manufacturing centre, Ontario's industrial revolution was fuelled by low power rates and high levels of service and reliability. Abundant water power at Niagara, a generous measure of populism, and an openness to statist solutions were added to the mix. How it all happened would become the stuff of myth.

In the beginning the language was as evocative as the aura that surrounded the deafening cascade and the foreboding whirlpools. The Falls were already receiving tourists from around the world. The visitors gazed with awe at Niagara's mighty torrent.

A Toronto *Globe* Saturday special on "Power, Light and Heat" had it that Niagara symbolized the continuing struggle "for the possession of power—not the mere animal power of political rule, but the actual subjection of the forces of nature." It was 1905, and a titanic political battle for control of Niagara's electrical potential was heating up, along with the rhetoric. Adding a bit of alliterative nationalism about "the progress and prosperity of a powerful people," the paper told its readers that The Falls required "the attention of all who are joined to mankind in the subjugation and enjoyment of nature." If the huge hydroelectric potential were to be realized, it would amount to nothing less than a "conquest."

A hundred years later the text deconstructors and discourse analysts would have had a field day. Environmentalists would identify this sort of talk with the worldview that gave rise to DDT and species extinction, with the ravishing of the Earth that accompanied untrammelled industrialization. During the sooty brown decades of the nineteenth century, when steam locomotives and smoke-belching factories stood as a shorthand for progress, coal had been the main source of heat and power. Niagara promised abundant supplies of clean, "white coal"—hydroelectricity. In the same way that nuclear power boosters would eventually claim that their projects offered a happy alternative to greenhouse gases, hydro backers at the turn of the last century could point to modern cities with bright lights and clean factories—a smokeless, electric future.

Ontario's manufacturing boom was lagging behind places like the English midlands and the Ruhr in Germany, where local capital and local coal spurred new industrial centres. U.S. industry was tearing ahead in the Allegheny–Great Lakes region. Nearby supplies of coal were essential because steam power functioned most efficiently when the factories were bunched together. Ready access to human energy was crucial.

Ontario's capitalists generally had sufficient labour, but their steam-driven plants depended on the anthracite coal miners of Pennsylvania. In 1902 nearly 150,000 Pennsylvania miners began a strike that threatened Ontario with a coal famine. As winter approached, the state militia was occupying the coal fields and the strikers were showing no sign of relenting. An arbitration agreement, drafted on the yacht of financier J.P. Morgan, eventually ended the strike with a partial victory for the workers.

Coal flowed once again, but in Ontario the lesson of dependence on U.S. energy was not lost. Niagara's height of land and its volume of water made it a natural place for hydroelectric generation, but in its ten years of holding the monopoly the U.S.-owned company that had been granted the charter to develop power on the Canadian side had taken no steps to do so. The company was producing electricity across the river, where new electro-chemical industries had already opened up to take advantage of cheap electricity, but on the Canadian side—and particularly along Ontario's Welland Canal, an ideal spot to locate factories—things were still quiet.

In the winter of 1903 a group of Toronto financiers set out to remedy the situation. The financiers, associated with the Bank of Commerce, the Toronto Electric Light and Toronto Street Railway companies, National Trust, and Canadian General Electric, obtained a Niagara concession for their newly formed Electrical Development Company. These "big interests," as they were known in the populist rhetoric of the day, recognized Niagara as a ready source of cheap power for their Toronto operations.

At the same time sixty-seven Southwestern Ontario mayors and factory owners (they were often one and the same) from cities such as London, Berlin (now Kitchener), Hespeler, Guelph, and Waterloo got together at the Berlin YMCA. Unlike their Peterborough counterparts, who already enjoyed cheap electricity from the Otonabee, these men had no access to the wonderful new energy. They had long been agitating for development of Niagara power, couching their arguments with reference to white coal. Among them was a cigar-box manufacturer who had twice been elected mayor of London, the largest city in Southwestern Ontario. Adam Beck was, in the words of one Hydro history, "a colourful personality," and he was about to launch a personal crusade that would dominate the rest of his remarkable career and establish him as a larger-than-life figure in Ontario political lore, a man with the same mythical status as Edison. If one image captured the mood at the YMCA that day, it was a cartoon of an eccentric Conservative politician named Beatty Nesbitt shaking an angry fist and demanding "The Government at the switch, not corporations."

The Berlin Convention set the stage for a bitter dust-up pitting the merchant capital of well-connected Toronto financiers against manufacturing interests based outside the provincial metropolis. Although the Toronto bankers and railroad magnates were never inclined to turn down government assistance, in this case the idea of the government operating a lucrative new business on a non-profit basis stuck in their ideological craw. They were not interested in losing out on a big money-maker. For their part, Beck and the troops in his public power movement regarded Toronto with suspicion, reasoning that if private capital got a chokehold on Niagara power, the needs of the street railway and electricity company would take precedence over their factories and towns.

The Toronto businessmen backing the Electrical Development Company were already busy with the "traction boom" in South America, where they were setting up street railways and the associated thermal and hydroelectric generating stations. Brazilian, Argentinian, and Mexican railway franchises also provided them with monopolies in electric power service. South American "traction bonds" were valuable securities, with the interest alone totalling more than a million dollars annually. The value of shares in Sao Paulo Tramway, Light and Power, placed at $50 in 1902, would triple within three years. The Toronto interests had a similar scheme in mind for a private Ontario Hydro.

In South America their syndicate would finance construction using bonds that the backers sold to themselves on a cut-rate basis. Bonus stock was thrown in to sweeten the deal. They would then place another, larger bonds issue, together with more bonus stock, with English investment houses. Organized insider trading would inflate the price of the South American paper by boosting interest in the enterprise. When broader public confidence was established, the syndicate would slowly start to dump its bonds and stock onto the market that it had created, turning a fat profit in the process. According to H.V. Nelles's definitive history of Hydro's origins, "Profits from these transfers, plus interest on the securities held, allowed the members of the syndicate to retain substantial holdings without having to pay any more than the initial margin. Of course, on a successful venture, the insiders' profits would be enormous."

The Toronto businessmen carrying out these Southern adventures so successfully were now aghast at uppity small-town capitalists who would have the government horn in on similar opportunities in their own backyards. Pork packer Joseph Flavelle, one of the most powerful men in Canada and chairman of the Bank of Commerce (and a proud son of Peterborough), apparently "spoke deprecatingly of Adam Beck." The tropical operations of the Toronto interests depended on supine politicians. Their investments in the South, explained B.E. Walker, general manager of the Bank of Commerce, were "based on charters given by countries very anxious indeed to have development take place and not having socialist views or envy of those who make money out of such charters." These men were admirers of the Liberal premier George William Ross, a model of fiscal probity. Ross was determined that the province not get into any debt over hydro development, and was open to control of electricity by the private sector.

Unfortunately for the Toronto interests, Ross lost out in the election of 1905 to James Whitney and the Conservatives, who swept the province by taking sixty-nine of seventy-six seats. Whitney, no social-ist, had campaigned on the promise of cheap public power from Niagara, and as soon as he was elected he cancelled a contract signed during the dying days of the Ross administration; it had granted to the Electric Development Company the remaining water rights at Niagara. Whitney sounded what must to the Toronto financiers have seemed a dangerous note: "I say on behalf of the government that water power all over the country should not in the future be made the sport and prey of capitalists and shall not be treated as anything else but a valuable asset of the people of Ontario." The plutocrats were choking on their port in the mansions on the gaslit streets of Toronto's new suburb of Rosedale. No one galled them more than the newly elected Member for London, the imperious Adam Beck.

Whitney appointed Adam Beck as Minister without Portfolio, but everyone knew that the outspoken booster of state ownership was the de facto Minister with the Power to Bring Power to the People. Within five months of taking office the government created the Hydro-Electric Power Commission of the Province of Ontario, which would assess the available hydroelectric resources and recom-mend the best way of distributing the power that would soon flow from them. Beck became chairman of the Commission, and soon after taking office he was on the road, selling his vision of cheap, abundant, clean power to enthusiastic public meetings. In rural areas he talked of electric farm machinery. In cities he delivered the promise not just of lights, but of electrical appliances in the home. Aside from "power at cost," Beck had a message that would long echo across the province: *the greater the demand, the cheaper the power.*

The Toronto financial bourgeoisie warned of a capital strike if Ontario were rash enough to go it alone and develop an electricity system without their participation. Henry Pellatt, the Upper Canada College boy who would go on to build the eccentric Casa Loma, predicted that government action would so frighten the cautious men of Threadneedle Street that London's capital markets would scorn Ontario bond offerings. He and the other Electric Development Company principals retained an advertising agency to purchase favourable editorial opinion.

Even though he was a prominent cabinet minister, Beck distrusted his own Conservative Party—certain elements in it regarded him as a dangerous fanatic—and bypassed Queen's Park in his public power crusade. A referendum was in order. The movement, with its base in municipal councils, drew up a standard public power bylaw and submitted it to the ratepayers in the local elections of 1907. The *Monetary Times*, the house organ of big capital, summed up the prevailing political dynamic of the day: "The very fact that the power companies seemed to be working so strenuously against the by-law must have cast thousands of votes on its side." The Toronto *World* warned that "the pure white light generated by God's greatest masterpiece, Niagara Falls" should not fall into the hands of middlemen like Pellatt. Every citizen, "however humble," should be able to have electricity at home, at cost. "We want no electric barons here as we now have coal barons of Pennsylvania, tolling us and tithing us." Voters from Ottawa to St. Thomas, inflamed by the florid language and the basic appeal to self-interest, passed the bylaw and were even joined in doing so by ratepayers in Toronto.

In 1908 the new Hydro-Electric Commission signed a contract to build a major transmission line from Niagara Falls to Toronto, with another line as well to the power-hungry cities in the Southwest. It was a victory for Beck, who insisted that cheap power was vital to Ontario's "supremacy" as an industrial centre. No longer would the citizenry depend on "a foreign nation for our coal supply."

The seeds planted by Adam Beck would flourish. Ontario would emerge as Canada's manufacturing centre, with Hydro as its sparkplug. Beck's distrust of politicians—even though he was a former mayor and a long-time cabinet minister—led him to do everything possible to insulate his company from their influence, and it would remain very much his company until his death in 1925. He denounced the government of the early 1920s because he thought it threatened to reduce Hydro to the level of "a government department debauched in politics." When he knew he didn't have long left, he instructed one of his minions to "Watch what they do when I am gone."

So great was Beck's power that soon after its election in 1919 this same government—a Farmer-Labour coalition—had even offered the Conservative the job of premier. But Beck had other, more important things on his mind, including the construction of the world's largest hydroelectric generating station at Niagara. It would eventually bear his name.

♨

By the early 1920s electricity had entered the homes of the urban middle class and was commonplace in the factories that had flourished during the war boom. The service also expanded rapidly as Hydro got busy wiring up the countryside. Hydro's new customers were suddenly part of a huge system. The power had its watery origins at Niagara or some smaller cascade. The complex network had at its centre the engineer, a man who could design a mammoth generator and make sure the power it produced found its way to bedside lamps and factory motors. From the earliest days, cost was no object—Hydro's first mega-generator was a case study in escalating costs—and from the beginning critics raised questions about Hydro's accountability. Darlington's critics would take note fifty years on.

Adam Beck first proposed the Queenston-Chippewa project in the middle of World War I, when power demand for war production was escalating rapidly. Locating the generator at the crest of the Niagara Escarpment, downriver from the Falls, would allow Hydro to use the greatest drop in height between Lake Erie and Lake Ontario. The choice sight, not far from the memorial to General Brock and the 1812 Battle of Queenston Heights, could take advantage of a drop that is almost double that of the Falls itself. The project involved massive tunnelling, the reversal of the flow of the Welland River, and a huge canal. Beck initially told the government that the cost would be $20 million, a huge sum in those days. When the predictable clamour arose, the government put the issue to a municipal plebiscite, but Beck was so confident of the outcome that he had already issued specifications to manufacturers of heavy generators and transformers. He took his engineers on a tour of the province, making persuasive speeches to specially convened meetings. The project was duly approved on New Year's Day, 1917. Planned "on a scale of the great European cathedrals," Canada's "temple of energy" went ahead, with the massive generators built by CGE in Peterborough. Later that year

Hydro got into the generating business in a big way, buying out its principal supplier, the Ontario Power Company, along with its water rights. By 1920, with the signing of a "Clean-up Deal," Beck's eighteen-year war with the Toronto syndicate was concluded. Hydro bought out its last major competitor.

In the early summer of 1920 Beck arrived at the office of E.C. Drury, the Farmer-Labour premier who had offered the power magnate the province's top job. Beck, who was described by his sympathetic biographer as "anything but pleasant in a number of his personal contacts," came right to the point, dispensing with the usual honorific niceties.

"Drury," he said. "This thing's running away with us, and I don't know where it's going to end. I think we should get some outside advice."

The premier knew that Hydro had done substantial estimate padding. Drury's auditors had also told him that although the legislature had voted $11,075,000 for construction of the magisterial power house and the canal, Hydro had already spent $14,713,970. Drury, who grew up watching thrifty farm women making soft soap from leftover cooking fat and hardwood ashes, found himself signing treasury warrants at the rate of a million dollars a week. He heartily agreed with the chairman's suggestion that they bring in the eminent U.S. hydroelectric engineer Hugh Cooper, who had just been hired by Lenin's Bolsheviks to work on the Dnieper River development.

A month later Beck again bustled into the premier's office. Apparently Cooper had identified soaring cost overruns.

"Drury, we'll have to let this fellow go. He's tied up with the private power interests."

If the employing class saw a Bolshie under every bed, Beck was obsessed by the Big Interests that had originally tried to stymie his public power plans.

"He saw their agents under every gooseberry bush," Drury would later recall. He allowed Cooper to continue. In the end the engineer's concerns proved well-founded. Queenston-Chippewa would cost $80 million, four times the original estimate.

Beck's hubris served the Hydro utility well in the years before and after his death, as the economic boom of the 1920s pushed demand for power. From its beginnings the Ontario Hydro-Electric Power Commission, as it was formally known until 1973, was pledged to supply cheap, reliable power at cost, and it did. This was not as easy as it might have seemed to city dwellers turning away from gas light and rural electricity users putting aside their coal-oil lamps. Hydro engineers were obliged to have sufficient capacity ready to meet the maximum possible demand, or "peak load," that would be placed on their central-station generators should everyone turn on the lights at once. It is a great truism that electricity is a commodity like no other because it cannot be stored. The physics of the inter-connected grid of generators, wires, and transformers means that the stable operation of what has been called the biggest machine ever made requires an exact match between the amount of power being generated and the amount being used at any one time. The peak load from household lighting demand occurs between dusk and bedtime, which was not in itself enough to justify the incredible capital expenditures of an operation like Queenston-Chippewa. How could that cost be justified?

The answer had been obvious to businessmen and engineers since the time of Edison: load-building. It was a time when electric traction motors were replacing horse-drawn trams. Electric motors powering machine tools and conveyor belts in factories had replaced steam generated from coal. Centrally generated electricity from Niagara was a boon to manufacturers with no access to handy mill-races like the ones that powered Peterborough's sawmills. Power flowing along new high-voltage transmission lines was boosting industrial production in small cities like St. Thomas and Kitchener, where Ontario's public power movement had its political base. Supplying power to factories was the perfect way of building load and amortizing the cost of pricey power stations and the transmission grid. Factories working two or even three shifts kept their motors running, furnishing central-station systems with a high and steady load. We have no way of knowing whether Beck consulted a crystal ball in 1914 before declaring to seven hundred municipal delegates at a public power meeting in Stratford, "Nothing is too big for us. Nothing is too expensive to imagine."

Hydro would spend two generations proving its founder correct, but from its earliest days the utility would have trouble coping with Ontario's appetite for power. During the 1920s it had to keep up with rising demand. By the time the Great Depression of the 1930s started, Canada had another new industry: thirteen thousand workers, almost all in Ontario, were building cars, with many more making parts and doing repairs. Another load-building technique was rural electrification. Hydro had strung its first rural lines in 1912. The Farmer-Labour government of the early 1920s offered big grants, subsidizing fully half the costs of transmission lines to the countryside. When the Conservatives returned to power in 1923 they recognized the value of extending electricity to the smallest hamlets and farms that dotted the back-concession roads of vote-rich rural Ontario. George Howard Ferguson's Tories soon extended the previous administration's grants to cover transformers and secondary equipment. Hydro had 1,205 miles of primary rural line in 1924, serving 20,605 customers. Some four years later it had 3,790 miles and 31,000 customers. The year before the Depression hit, it added 929 miles and 6,000 customers.

Hydro engineers doubled as salesmen, convening meetings and telling prospective customers that the more people who signed up per mile of line and the more power they used, the cheaper that power would be. One Hydro engineer went out of his way to sign up a shrewd poultry farmer, who happened to be the third potential customer on a line. He agreed to buy thirty dozen eggs a month from her, which would help her offset the costs of lighting her house and chicken coops. The engineer found himself retailing eggs until the farmer, realizing that cheap power was a godsend, let him off the hook.

By 1926 Ontario's rate to domestic consumers was just under 2 cents per kilowatt-hour, while the fee for comparable service in the United States, with its private-power, laissez-faire system, was 7.4 cents. In 1935 inexpensive electricity was still beyond the reach of millions of Americans, and only one in ten farms had access to central-station power. By 1945, 42 per cent of Southern Ontario farms already had access to "the juice," and it would take a combination of the New Deal and the postwar boom for rural electrification in the United States to catch up with the pioneering efforts of Ontario, which also led the way in Canada. According to Hydro historian Keith Fleming, "Until the 1930s, no other utility anywhere in North America came close to matching the resources that Hydro poured into developing a rural market. Certainly, there would be little doubt in the minds of the approximately 154,000 residents with rural service contracts at the end of 1945 that public ownership had paid off." It would not be until the mid-1950s that 96 per cent of rural U.S. homes would have access to electricity.

Within a year of Beck's death, Hydro engineers were predicting that the capacity of the Niagara power plants would fall short of demand by 1928, with the gap increasing eightfold by 1933. Ontario began signing contracts with private power producers in Quebec. But for the ambitious planners, that was hardly enough. The engineers had plans to "go back to Niagara" to meet the 11 per cent annual growth that Hydro was forecasting for the 1930s, even if it meant turning the Hudson Bay watershed around by diverting north-flowing rivers southward into the Great Lakes system. This, they figured, would increase the flow of water over the Falls—or through their nearby penstocks. This wacky plan stemmed in part from the usual technological hubris; the idea of taming this or that watershed was by this time commonplace at Hydro headquarters. Hydro had access to almost unlimited credit and a growing corps of engineer-managers eager to deploy the funds.

But aside from the growing can-do mentality and the expansion in the use of electric power, there were two reasons for the expansion imperative—reasons built into the public system that Ontario had established. If things went wrong and supply got disrupted or the price spiked, the government would be left standing alone to take the heat. There would be no private monopolist to blame. Six months before the great crash of 1929, Beck's successor at Hydro explained the political calculus to Premier Howard Ferguson: politicians presiding over power disruptions did so at their peril. "Public ownership in the electric power field," Hydro boss Charles Magrath said, "will always differ from private ownership in one main feature, that is the former must always have ample supplies of power available, and must be prepared to go further than a private corporation could reasonably be called upon to attempt in meeting a demand for power."

Then there was what T.H. Hogg, Hydro's long-time chief engineer and later chair, called "the promotional rate structure." Hogg, described in Hydro's semi-official history as one of the utility's "great engineers," explained that the rate structure meant that the more power you bought, the less you paid. Hogg knew how long it took to move a megaproject from drawing board to ribbon-cutting and secretly promoted the huge St. Lawrence Seaway hydro scheme even when Mitchell Hepburn, the mercurial premier of the day, was opposing it during one of his angry spats with Ottawa. In 1941 Hogg offered a concise explanation of the big-is-beautiful mindset that would persist for half a century or so—at least until after the ribbon at Darlington was finally cut.

> The basic idea behind the promotional rate structure is this: the greater the load density on an electric distribution system the greater the economy of operation and use of materials; the larger the demand for power the greater the opportunity of developing large power resources and the greater the economies that come from generating on such a large scale. These factors lower the cost of power to consumers.

The Seaway megaproject would have to wait until the 1950s, when very different conditions prevailed. In the meantime Hydro set up a Department of Sales Promotion. Unveiled in the 1938 annual report, the department would work with municipal utilities and appliance manufacturers to make sure that the low-cost power would find an outlet. This early version of "Live Better Electrically" would be temporarily shelved in favour of wartime conservation campaigns: demand-side management would replace the load-building strategy of the promotional rate structure. Electricity was suddenly a vital "war weapon" aimed at the production of tanks and mortars. But Hydro's end-of-Depression thinking and sudden shift to wartime austerity underlined two other aspects of the electricity equation: it would always be difficult to predict power demand and balance demand with supply. Still, although Hydro proved during the war that it was able to conserve power by actively discouraging demand, it could just as easily pull out all the stops to get people to buy, buy,

buy—and that is exactly what the utility did for the thirty long years of the postwar boom.

⊕

The gizmo is as commonplace as the mailboxes beside the front door. Every house has one. It is so ordinary that, like the electrons it measures, the little glass cylinder gets taken for granted. In the electricity system the meter serves as the border, the dividing line between the utility operating the grid and its customers. The "grid"—generators, transformers, switches, and wires—is handled by the utilities. Lights, refrigerators, televisions, and garage-door motors are the responsibility of customers. When you turn on an electric heater on a cold winter night or crank up the air conditioning during a heatwave, the little metal disk behind the glass in the meter reacts. Instead of turning like a lazy little merry-go-round, it goes into a frantic whirl.

Until well into the 1970s, people on the customer side of Ontario's meters paid little attention to what was happening on the utility's end of things. The lights invariably went on at a flick of the switch. There was, apparently, an easy abundance of supply, and most people were not concerned at all about how fast the little disk was spinning. The price was low. Electricity remained a minor item in household budgets. Things hadn't changed in the twenty years since Hydro had hired W.R. Harmer from Peterborough's Canadian General Electric to put the Promotional Rate Structure permanently in place with the "Live Better Electrically" campaign. The ads featured a smiling bride brandishing an electric eggbeater—"the gift that really helps her to live happily ever after!" The Sister of the Skillet who baked the best cherry pie got a new electric stove. When it came to the customer's meter, Ontario Hydro wanted to keep the disk spinning faster.

Yet when systems engineer Bob Lake came to Peterborough in 1986 to head up the Electric City's municipal utility, the Public Utilities Commission had not been playing the load-building game. Instead it was still practising wartime-style "load-shaving." Lake found that the PUC was using high-frequency signals to turn off its customers' water heaters during periods of peak demand and back on again when pressure on the system dropped. The approach had initially been a way of conserving power for the war effort. Keeping the wartime production lines humming at CGE and other factories was more important than having a hot bath. But even later, during the heyday of Hydro's postwar "Live Better Electrically" campaign, Peterborough continued to control power use on the other side of the meter.

Lake explained it as a matter of simple common sense. "The local utility was really well placed to do this sort of thing," Lake said. "Since the utilities were not for profit, all customers benefited." Few customers noticed the difference in the supply of power. Power use tends to rise sharply in the late afternoon as people start to use appliances. If you turn off the power to thousands of water heaters during the peak supper-hour period—the electric rush hour—both the customer and the local utility save money. Lake, a system engineer who was already a fourteen-year veteran of Ontario Hydro's generation and customer service operations when he moved to Peterborough, knew immediately that this load-shaving flew in the face of his former employer's attitude to power use. The program was so successful that the Peterborough distributor was able to offer free hot-water tanks to customers. The year after Lake took over at the PUC, the utility removed thousands of round tanks from basements and replaced them with larger, 60-gallon water heaters. Each one had a little box labelled "Energy Miser Plus."

Lake had started Ontario Hydro's Lakeview generating station in Toronto and eventually ran the system in the company's biggest geographical area. A cost-conscious manager, he knew all about the favourable contracts that the Power Workers' Union had negotiated with Hydro, which was doing virtually everything with its own staff. When he arrived in Peterborough it was costing the PUC $1.25 to read a meter. He eventually reduced that to 35 cents by contracting out the work to the private sector. Not long after moving to town Lake learned of four unused dam sites on the Otonabee River upstream from Peterborough. He saw "water spilling over those dams all day, every day, and that's wasted energy." He figured it would be a good idea for the PUC to use those dams to produce its own power, thus cutting costs and saving on purchases from the Pickering nuclear station and the controversial Darlington plant, then still under construction. It was not to be. It was easier to face down the union than Ontario Hydro.

"We wanted to build generation but Hydro said, 'No. We are not going to approve your capital expenditure budget because we don't want competition.'" From Lake's point of view, Hydro "let all that renewable energy go to waste."

Other PUC managers in the area were also upset by how Ontario Hydro was acting in its dealings with their companies and the environment. For nearly seventy-five years a power house on the Trent River had provided Campbellford, southeast of Peterborough, with cheap, reliable power: the citizens there enjoyed the lowest electricity bills in Ontario. In 1981 the turbine shaft cracked, rendering the power plant useless. Campbellford found itself buying power from Hydro at a higher cost and turned to the provincial Energy Ministry for a low-interest loan to get its own hydro power back. The Ministry, an outfit well below Hydro on the political pecking order at Queen's Park, refused to help. Although, as it said, it supported small-scale hydro "in a general sense," it would not lend the money. With interest charges that equalled the cost of a new turbine, Campbellford's PUC was forced to pay roughly double the price. Still, the town's investment "in a renewable, inflation-proof source of power" paid off, keeping the rates lower than Hydro's.

"Hydro wants to control everything. They always have and they always will," the Campbellford manager complained at the time. That same principle was brought home to Lake when Hydro officials came to Peterborough to look into the local utility's conservation programs. By this time Peterborough had the biggest load management program in Ontario, with some twelve thousand basement water heaters being regulated on a short-term basis from the central control room.

"We put in 60-gallon tanks and were controlling them for four to five hours," Lake recalled. But Hydro said, "No, that's not good enough. You've got to have sixteen-hour control or it's not worth anything." According to Lake, "In typical Ontario Hydro fashion, they couldn't be satisfied with twenty-five cents. They had to have a dollar."

The grumbling of managers at places like Peterborough and Campbellford typified the growing unease about Hydro. The boss of the Peterborough PUC was neither anti-nuke nor a small-is-beautiful wind-power activist. Lake was an engineer who wore the trademark iron ring. Having learned the electricity business inside Hydro, he understood the utility's culture, the mindset of the men who made the decisions. He was becoming increasingly aware that the company's methods often clashed with what he saw as more efficient ways of dealing with Ontario's future electricity needs—and, indeed, with the corporation's prospects for financial health.

The various inquiries during Bill Davis's minority governments (1975–81) shone a glaring public light on Hydro's planning and priorities for the first time in history. The results, at least from Hydro's point of view, were not encouraging. Dr. Arthur Porter's Royal Commission report had offered a polite but firm indictment of Hydro's expansion plans. But now there was also another factor at play: the local utilities, which had been key Hydro supporters since the days of Adam Beck's public power movement, were starting to become disaffected by the company's high-handed attitude and indifference to their concerns.

Although Davis was a master at keeping the opposition off balance and staying in office, he was powerless to avoid the creeping sense of suspicion surrounding Hydro. Although notions of transparency and accountability had not yet made their way into common political parlance, that is what Hydro's critics were calling for. All of which gave the environmentalists who formed the backbone of Hydro opposition some small solace. But the apparent shift in the wind still did nothing to change Hydro priorities once the Conservatives, promising to renew Ontario's pricey nuclear future, were re-elected with a solid majority late in the winter of 1981.

Some three years later, after thirteen years as premier, Davis surprised everyone by announcing that he was about to step aside. In one of his final official acts as premier, he got out the scissors one more time to preside over the opening of the Darlington Information Centre. He surprised no one in his brief remarks at Darlington.

"I have always been an unabashed supporter of Ontario Hydro," the genial lawyer intoned. "I can think of very few things that have given me greater satisfaction than to see the way Ontario Hydro has grown."

Indeed, the company—and its debt—had grown considerably just because the Conservatives treated it like a favourite son who routinely gets indulged with meals at fancy restaurants and never has to look at the right-hand side of the menu. The Darlington Information Centre alone cost $1.1 million. Another election was looming. Hydro's new chairman, Tom Campbell, chimed in with the sort of get-it-built talk that had always paid off on the Tory hustings. "Darlington is now the largest construction site in North America," boasted Campbell, who had just moved over from the deputy's job at Treasury to chair the utility. "When it is completed it will have eighteen times the amount of concrete that the CN Tower has."

Talk of cement-pouring megaprojects always played well with Ontario's gravel-gouging interests who made up what they liked to call the "aggregate industry," a key prop of Tory support in Southern Ontario. Campbell had been part of the transition team when Davis took over from John Robarts. He now mused that Darlington would rank with Toronto's proposed domed stadium as part of Davis's legacy.

All the pundits expected the Tories to coast to another victory in 1985. The economy was back on track after the deep recession. The Liberals under rookie leader David Peterson, who had said he favoured immediate cancellation of Darlington, were reeling from the defection of Sheila Copps, the only woman in the caucus, and Don Boudria, one of their two Francophones. Both of the promising political talents had left for federal politics. Peterson's party even trailed another untested leader, the NDP's Bob Rae. A quick-witted debater, Rae was an extreme moderate who believed instinctively that political success in Ontario lay straight up the middle of the political spectrum.

Davis had alienated many core Tory voters by extending full funding to Catholic high schools, and the convention that chose his successor, Treasurer Frank Miller, was described by one astute Tory seer as "peculiar." Dalton Camp noted that the Tories had emerged from the $5-million leadership race as a party divided between older and younger, city and country, left and right. "I am not sure that it was as beneficial to the party as these things usually are."

Along with separate-school funding, the environment emerged as the sleeper issue of the election campaign. When a truck spilled its toxic load beside the highway in Northwestern Ontario, Miller was slow to react and the opposition pounced. With the polls showing that voters were reacting to the usual time-for-a-change rhetoric, the Tories came up with a new program in the middle of the campaign. Alongside a promise that all students would have equal access to Catholic schools, three of the other ten belated promises dealt with environmental issues: hazardous waste transportation, landfill cleanup, and a war on acid rain. But it was all too late. Although Peterson's Liberals won fewer seats, they had more votes than the Tories, and a round of hectic horsetrading began. The Liberals dropped hints about former NDP leader and Hydro gadfly Donald MacDonald being named Hydro chairman. The Tories told the NDP that they would at last move to cut acid rain emissions from Hydro and Inco. In the end, a Liberal-NDP accord promising no election for two years gave David Peterson the government.

Hydro affairs had not been going smoothly. The nuclear critics had been maintaining a grim, we-told-you-so watch since 12:10 p.m. on August 1, 1983. That Monday, while most people were relaxing on the last lazy day of the midsummer long weekend, technicians at the Pickering nuclear station scrambled frantically to contain the first loss of coolant accident to hit a Candu reactor. A two-metre rupture in pressure tube G16 produced a leak that initially pumped a ton of heavy water per minute onto the reactor floor. It took two weeks to stop the leak, which cost Hydro a quarter of a million dollars a day in replacement coal power. For years advocates of a soft energy path had been describing nuclear power as "brittle." They said that unlike wind and solar technologies, which had never enjoyed the billions in government subsidies lavished on the Candu, Ontario's nukes were vulnerable to dangerous and costly mishaps that might shut down entire stations. Now that just such an accident had occurred, it emerged that the zirconium alloy pressure tubes were subject to "hydriding," a phenomenon well known to metallurgists. The alloy absorbed hydrogen atoms, making it vulnerable to cracking: brittle.

Re-tubing Hydro's reactors promised to be expensive and time-consuming. Ontario's nuclear future was starting to look dim—and equally expensive. "The cost of re-tubing Pickering A and Bruce A alone has been estimated at $350 million, no doubt a conservative guess," concluded the *Nuclear Free Press*, an energy watchdog paper published by Peterborough greens. "The embrittlement problem also heralds a new era in which problems associated with reactors can no longer be written off as the teething problems of a young technology. Now we're already witnessing hardening of the arteries—a disease of old age."

By the time the Peterson government took office the Pickering debacle had temporarily faded into the background. The main issue was not Hydro's costly nuclear breakdown or its sulphurous acid rain emissions, but Darlington. The Liberals were now content to buy time. Their accord with the NDP included a pledge to set up yet another legislative committee to examine Hydro affairs. Indeed, for the NDP hierarchy, opposition to the huge, expensive nuclear project had never been completely front and centre, in part because CUPE Local 1000, the union representing over twelve thousand Hydro workers was, not surprisingly, adamantly pro-nuke.

In the mid-1980s Hydro's supply-demand equation was simple enough. Even without Darlington's 3,500 megawatts, the utility had more than enough capacity to meet peak demand. Hydro could put out 27,300 MW, some 40 per cent above peak demand. The company was still in the clouds as far as its demand forecasts were concerned. It was predicting that it would need 30,000 MW of peak power by 2000, and that was still not factoring in significant conservation and energy-efficiency programs. In late 1985 the Select Committee on Energy recommended that Hydro be allowed to complete units 1 and 2 at Darlington, pending its final report. Brian Charlton and Ruth Grier of the NDP opposed the proposal, arguing that the money and effort being invested in the nuclear plant were deflecting Hydro from the more important priority of investing in renewable energy and conservation.

By the time the committee published its final report, the nuclear power industry looked like it was down once again. In April 1986 the Soviet reactor at Chernobyl in the Ukraine blew up, sending a cloud of radioactive poison across the landscape. In the first week thirty-one people died from radiation exposure and eventually over 200,000 people were evacuated from a 4,300-square-kilometre area. Agriculture and forestry in the area had to be abandoned.

If the industry was down, it was not out. The 1986 final report of the legislative committee read like a reprint of the Porter Commission's indictment of Hydro's manic fixation on nuclear expansion at the expense of conservation. It proposed small-scale generation coupled with more investments in energy efficiency. But even though Hydro was now estimating the final price for Darlington at $11 billion, the committee's advice to the Peterson government echoed that offered by the Victorian mother on her daughter's wedding night: "Lie back and think of England . . ." There was nothing to do but grin and bear it—Hydro would be allowed to complete the station. The utility had already spent $7 billion. Once again the NDP members dissented with their good-money-after-bad argument.

The greens were frustrated. They pointed to a tragic pattern of Hydro underestimating project costs. They imported the obligatory U.S. expert Amory Lovins to testify that cancelling Darlington would allow Ontario to get more bang for its energy buck by redirecting spending to efficiency. "Finishing Darlington was a renunciation of the basic economic principle that sunk dollars don't count," recalled lawyer David Poch, who was working for Energy Probe at the time. "The only question is how do you most efficiently spend the money you're going to spend."

The Liberals refused to halt construction at Darlington pending their final decision, thereby digging themselves ever deeper into a political hole as Hydro kept on spending. "I had a hell of a problem with Darlington," David Peterson said later, looking back at the *realpolitik* of the affair. "We had said we wouldn't have started Darlington. Then, the more you're digging . . ."

There were also the facts of life of minority government to consider. Peterson believed that his 1985–87 minority was much too strong to risk sinking it on an issue like Darlington. Besides, the NDP would never have considered breaking the accord by pulling the plug on Peterson over a nuclear construction project that was not an issue on a public radar screen dominated by the question of extra-billing by doctors. In the event, the cabinet gave Darlington the go-ahead before the 1987 election that swept the Liberal government to an overwhelming majority. Peterson thought that he had no choice but to approve Darlington and to hope that its price tag would not rise above $11 billion. "My view was when you've spent seven billion dollars, you have to look at it and ask what the alternatives are, what are the needs. You have to look at supply. . . . We came to the conclusion it was more responsible to proceed than not to proceed."

One of the most powerful figures in the government was Peterson's attorney-general, Ian Scott, a canny reforming lawyer who quickly learned not to trust Ontario Hydro's spending estimates. Scott later reckoned that Darlington's interest bill alone totalled more than Ontario's investment in the SkyDome, the symbol of the boom-time spending of the 1980s. He put the decision to approve Darlington at the top of his list of the seven biggest challenges the Liberals faced in their first term. "This was," he later recalled, "undoubtedly the worst decision the government ever made."

Although green activists soured on the Liberals for what they saw as a cave-in on Darlington, the five years of Peterson's government were marked by a pronounced shift away from the Davis days. Environment Minister Jim Bradley ordered Hydro, Inco, and other high-stack polluters to make major cuts to sulphur dioxide emissions, and he proclaimed a spills bill that made polluters liable for toxic mishaps. Scott was suspicious enough about Hydro that he made sure the utility was subject to the provisions of his new Freedom of Information Act. This move opened the utility to even greater public scrutiny.

There were other signs that things were changing. The Peterson government appointed the first non-engineer to sit in the president's chair: Robert Franklin, who arrived from CN Rail. Tom Campbell remained as chairman despite his Tory connections, but caused an uproar when he mused publicly that more nuclear plants were in order. The first throne speech from the majority government promised more public "accountability" from Hydro. Treasurer Bob Nixon's 1989 budget announced a major change in the Power Corporation Act: Hydro would be forced to make annual payments to the government in return for the debt guarantees that had been a prerequisite for the expansion plans still on the drawing board. The Liberals also amended the Act to expand the definition of conservation to include a reduction in electricity use, and they opened the door to the possibility of private generation. Henceforth Hydro would have to include the promotion of non-utility generating companies as part of its legal raison d'être. Owners of small hydro dams and those who wanted to erect windmills had long been embittered because it was virtually impossible to get access to the Hydro grid, let alone get a decent price for their power from the monopoly utility. Environmentalists and soft energy path boosters expressed optimism about the Ontario government's attempt to control Hydro and deal with matters green by getting serious about energy efficiency and renewables.

Developments elsewhere offered no such cheer. Immediately upon being elected in 1984, the federal government of Brian Mulroney had scrapped Canartech, the Crown corporation that invested in conservation and renewable energy projects, and eliminated the Canadian Home Insulation Program. The federal Tories had also dispensed with the National Research Council's $300 million energy research and development division, and with it a crucial Canadian wind and solar power initiative. Work on more energy-efficient windows also went out the door. Ottawa was investing its political capital in a continental energy strategy that would culminate in the Canada–U.S. Free Trade Agreement. Henceforth Canadian energy policy was bound by treaty obligation to a speed-up in the delivery of non-renewable resources to the U.S. market.

Although the FTA was a let-the-market decide initiative generally weighted against government intervention, it did permit Canadian subsidies for frontier oil and gas development. It prohibited not only differential pricing of those resources but also conservation measures that would stem the exports of Canadian energy in the event of a supply crunch. The bitter political dispute over free trade was played out against a broader background marked by the ascendance of what would become known as "neo-conservatism" in North America, "neo-liberalism" in much of the rest of the world.

four

Liberalizing Electricity: Chile, Britain, and Ontario

What went wrong in the England-Wales Power Pool system? Pious devotion to theories about market forces at first blinded the government to its obvious failings. . . . Trouble comes when regulators are enticed into believing that competitive markets, where none exists in reality, can substitute for public control.
— Greg Palast, Jerrold Oppenheim, and Theo MacGregor,
Democracy and Regulation: How the Public Can Govern Essential Services, 2003

ENVIRONMENTALISTS WERE not the only ones manoeuvring to take advantage of the visible cracks in what historian Richard Hirsh has called the "utility consensus." The sense of inevitability that surrounded the system of publicly owned (or privately owned but closely regulated) monopolies supplying electricity from large centralized plants was increasingly shaky. It wasn't just the small is beautiful crowd who were putting forward alternatives.

Most standard histories of electricity privatization begin with the decision by British prime minister Margaret Thatcher to embrace the superior efficiencies ostensibly inherent in free markets. But well before the Thatcherite experiments in selling off Britain's public power system, Chile's General Augusto Pinochet got into the act.

For many North Americans the bloody 1973 putsch that overthrew the elected government of the socialist Salvador Allende was just another Latin American coup. But Chile was no banana dictatorship. Chileans were proud of having the second-longest history of unbroken democratic government in the hemisphere and the highest standard of living in Latin America. A political shift had come about in the country in the presidential election of 1970, when a newly enfranchised working class and peasantry voted for a coalition, led by Allende, that promised to redistribute land and nationalize the foreign mining companies. Once in power, Allende set about doing just that, and his efforts did not go unnoticed. Three years later, on September 11, 1973, with the support of the CIA, General Pinochet's forces bombed the Presidential Palace, murdered Allende, and seized power.

Pinochet's ruling military junta had the support of the local oligarchy and the middle classes, as well as transnational corporations that now had the chance of recovering companies that had been nationalized under Allende. Thousands of leftists were simply "disappeared." The junta, as historian Eric Hobsbawn put it, "introduced Chile to the characteristic features of 1970s military regimes—executions or massacres, official and para-official, systematic torture of prisoners, and the mass exile of political opponents." It also introduced the country to economic liberalism.

While traditional Latin American military strongmen either tried to use the state to boost domestic industry or line their own pockets, the Chilean generals handed economic management over to a coterie of economists who, in the words of one of the ruling generals, "agreed with each other and gave us simple answers to our questions." Since the 1950s, four or five bright young economics students from the Universidad Católica in Santiago would head north every year to graduate studies at the University of Chicago, a haven for classical, laissez-faire economics during a time when Keynesianism dominated most scholarly and policy work. Learning their lessons from professor Milton Friedman and his disciples (who would go on to advise U.S. president Ronald Reagan), most of the Chilean students absorbed an almost messianic zeal for reducing the role of government and letting markets rule. When these so-called "Chicago boys" returned to Chile, they spread the good word of the invisible hand of the market. In 1973, when the iron fist seized power, a hundred highly trained, free-market zealots were ready to take over economic policy.

Long before the International Monetary Fund made such programs mandatory, the Chicago boys set about slashing social spending, freeing prices to be set by the market, reorienting the economy to trade with the rest of the world, and selling off state-owned enterprises. Between 1975 and 1989 they sold government stakes in more than 160 corporations, 16 banks, and over 3,600 agro-industrial plants, mines, and real estate. Many of those enterprises went bankrupt and had to be subsequently renationalized, but the electricity sector proved to be remarkably profitable.

For the Chilean generals, privatization was part of a broader strategy to reduce the role of the state and enhance the power of markets so that the socialists could never again gain a toehold in the economy. They introduced electricity privatization and deregulation legislation in 1982. Some twenty years later the U.S. Department of Energy was still calling it "a shining example of how free trade policies and privatization efforts can fuel economic growth. . . . Chile continues to be the model for privatizing the electricity industry and unbundling the generation, transmission, and distribution systems."

Chile clearly had the kind of "strong leadership" that free-market economists came to admire. According to Hernan Büüchi Buc, who oversaw the privatization of the electricity system as Chile's finance minister from 1985 to 1989, "The liberalization of the economy requires a strong visible leader, someone who is willing to absorb criticism and who, in addition to requirements in the area of technical competence, must also possess perseverance and an overriding faith in the market and free enterprise."

The largest single privatization was the giant state-owned electrical utility ENDESA, with a net worth equivalent to 7 per cent of Chile's Gross Domestic Product. Members of the armed forces got preferential access to ENDESA shares, creating a powerful vested interest that would oppose renationalization. The regime's administrators privatized all of Chile's major electrical utilities between 1986 and 1989, getting the complex task over with before they were forced from office and into lucrative positions with private-sector electricity firms.

In a harbinger of what was to come, though, the Chilean free-market dream did not lead to the kind of competitive market that its advocates envisaged. By the late 1990s the Chilean energy market consisted of only three major participants. The largest company owned 60 per cent of the installed generating capacity, owned outright the transmission and distribution facilities serving half of the consumers, and owned 70 per cent of the remaining water rights that could be potentially used to generate electricity. Not surprisingly, that company has not been shy about throwing its weight around. According to two Chilean economists, "Potential entrants are afraid of confronting this behemoth, given the possibility of discrimination within the pool, the lobbying power of the dominant firm, the problems in legislation, the possibility of discretion by the regulator, and the inefficiency of the judicial system for companies seeking redress."

The shareholders, not consumers, have reaped the benefits. Following privatization, the rate of return for Chilectra, the distribution division of ENDESA that served 40 per cent of the market, tripled from 10.4 per cent in 1988 to 35 per cent in 1997. It had become much easier for the private company to cut off people who weren't paying their bills. (In a sort of unofficial subsidy, poor people tapping into power lines in poor countries were generally ignored by state-owned electricity companies.) Time after time technocrats who oversaw privatization found their way into the executive suites of privatized corporations, where the much more generous private-sector salaries just might explain their interest in pushing privatization in the first place.

In Britain the resolute Margaret Thatcher was always so determined to ram through controversial measures that she became known as the Iron Lady. She said famously, "The lady is not for turning," and infamously that there is no such thing as society, only individuals and families. Well after she had been tossed aside by her own party, another of the former prime minister's nostrums lived on, symbolizing the dominant ideology of her time: "There is no alternative." Even if the initiative in question seemed ill-advised, very costly, or just plain foolish, it was enough to declare that there simply was no alternative.

Thatcher, who had tremendous admiration for Chile's Pinochet, also reflected the Chileans' high standards for a strong leader willing to persevere when confronted by firm opposition. Before becoming prime minister she had helped establish the Centre for Policy Studies, a think tank whose goal was to convert the Conservative Party to the now-familiar form of economic liberalism rooted in small government and unrestricted markets. The Centre, along with other like-minded think tanks and business groups, would later lobby her on electricity privatization. She was, not surprisingly, receptive.

When she was first elected prime minister in 1979, Thatcher's platform made no mention of electricity privatization, but by the time she captured her third mandate in 1987 it was a central part of her campaign. The government published a policy paper, "Privatizing Electricity," in 1988, and by April 1990 the publicly owned monopoly structure had been replaced with one based on private ownership, supposedly competitive markets, and independent regulation.

Electricity privatization was justified on the grounds of efficiency. Without the discipline of market forces, the think tanks argued, there was no incentive for publicly provided services to be efficient or responsive to consumers. Those who took up this cause were by no means stymied by the lack of hard evidence for this line of thought. To the contrary, studies in the United States had found that public utilities were as efficient as their private-sector counterparts and that public power was on average cheaper. Instead the advocates conveniently bypassed the question of evidence by basing their arguments on pure theory: markets are by definition more efficient; hence privatization must increase efficiency and reduce waste.

The rationale for pursuing the privatization enterprise in Britain was based, first of all, on an ideological quest: to reduce the role of government in the economy and limit the ability of future governments to intervene in the private sector. Related to this was the more pragmatic political goal of debt and tax reduction, because the proceeds from the sale of the utility could be used to pay down the debt and/or lower taxes, a short-term measure that meant passing up future revenue streams. Privatization could also be used to reduce the power of unions, always a central goal of market fundamentalists. By forcing British coal miners to compete with foreign coal and the new, cheap sources of natural gas in the North Sea, the government could break the hold of the powerful National Union of Mineworkers over electricity supply (with sixty-six thousand jobs lost in the electricity industry). Wages and working conditions deteriorated. Finally, privatization was part of Thatcher's vision of creating a shareholder democracy based in a "new breed of owners," with more Britons owning shares than holding union memberships. To this end, shares were made widely available and promoted through massive advertising campaigns.

British industry backed electricity privatization because it supported Thatcher's ideological project, shared her hatred of unions, and saw the possibility of lower prices for itself. Privatization could also be used as a form of cover for subordinating equity and environmental goals to economic objectives, and big power-users quickly grasped the opportunity to use their purchasing power to get the cheapest rates. Between 1988 and 2000, industrial consumers saw their prices drop by 30 per cent while prices for domestic users dropped by only 24 per cent.

Advocates of electricity privatization point to these price reductions as evidence that competition works; but there were even greater reductions in the United States, and even in Ontario prices declined in real terms from 1994 to 2003. Lower prices in Britain, it seems, were not a consequence of effective competition, but rather of lower fuel costs and massive layoffs. The price of coal dropped by 30 per cent, while cheap natural gas from the North Sea fields became available, dropping the price by 40 per cent. This had the welcome side effect of reducing air pollution, as cleaner natural gas replaced coal— a benefit that was not a consequence of privatization but of the development of the North Sea gas fields. During the same period, 46 per cent of the workers in the electricity sector lost their jobs, which lowered costs but raised concerns about system reliability and safety.

Indeed, without "competition" the price might have dropped even more, because the market-based electricity-pricing system adopted in Britain—which would become the model for restructuring around the world, including the United States and Ontario—has never worked very well. Or at least it has never behaved like the outline of competitive markets featured in textbooks would seem to indicate. If you happened to be selling electricity, it worked so well that it bordered on a licence to print money. National Power, one of the two privatized generation companies, paid out in dividends to its stockholders in a single year an amount of money that was worth more than the entire value of the company's stock at privatization.

The Power Pool bidding system was supposed to lower prices for electricity by subjecting power generation to competitive market forces. The theory was straightforward: every day generating companies would bid to supply the nation's grid system for each half-hour of the following day. The lowest bidders would supply the system, and consumers would benefit as ruthless competition between the owners of private power plants led to lower prices. Markets, not regulation, would set electricity prices.

On paper, it was a thing of beauty—a simple, yet elegant model that made economists feel like poets. In reality, it proved to be a costly fantasy. The utilities that sold the power to small consumers were prohibited from buying electricity through direct long-term contracts with generators, but rather were exposed to a highly volatile, and easily manipulated, short-term spot market in which commodities or foreign exchange are sold with lightning speed, for immediate delivery. Individual companies, however, could buy power directly through long-term contracts at stable prices. But onlookers in Britain soon found signs of market manipulation. A February 1999 study by the British agency that regulates the market concluded:

> There is strong evidence that manipulation of Pool prices has in fact been occurring; that participants in the Pool have been operating within the existing Pool Rules to take advantage of those rules for their commercial interests—the "gaming" of complex rules; that prices have been manipulated; and that higher wholesale prices have been established which will result in higher prices for consumers. The occurrence of this manipulation has been accelerating.

When the technocrats who run electricity markets use the word "gaming," they are politely referring to finding loopholes in the rules. What it amounts to is collusion, price gouging, lying—in short, cheating the consumer. Endless attempts to close the various loopholes have failed, as companies find new ways of rigging the bidding. Paradoxically, the frequent reworkings of the system have meant that there are now more regulations under the "free market" than there were under the government-owned and regulated system. Deregulation, a term often used to describe the whole process, becomes a gross misnomer.

All of this would probably be more amusing if this flawed system hadn't been taken so very seriously in North America. It came first to California, where "gaming" the market was elevated to a form of high art. Indeed, the same consultants who designed the markets for the government turned around and offered high-flying new energy companies advice on how to cheat.

🕮

By 1989 the idea of privatizing Ontario Hydro had found an ally in the offices of Toronto's Energy Probe. The former environmental group was re-imagining itself as a "consumer" group while imagining that the Pinochets and Thatchers of the world were on to a good thing with their sell-off of public power companies. Energy Probe had seized upon neo-conservative politics with all the righteous zeal of the recently converted. There was no way that Ontario Hydro would ever change its ways, Energy Probe argued, without a stiff dose of market discipline.

Over at Hydro tentative signs appeared of a shift in the no-expense-spared gigantism. Soon after Robert Franklin took over as president, Hydro's Marketing Branch was renamed the Energy Management Branch. Serious effort at power-saving programs would have presented a huge challenge for any utility, let alone one whose main business for seventy-five years had been the sale of more power to increasing numbers of customers. Hydro had to be persuaded that Ontario could realize a whopping total of 10,180 megawatts in demand-management savings by the turn of the century. This was in itself a megaproject, albeit of a different nature than the building of a four-reactor, 3,500-megawatt plant like Darlington. It would require both persuasion and the designing of a complex set of incentives, subsidies, and regulations. A diverse group of engineers and architects, builders and building managers, manufacturers and distributers of appliances, light bulbs, and motors would have to be enlisted along with the trade and professional associations that represented them. If successful, it would mean "building in" to factories, office buildings, large stores, hospitals, schools, and houses better, cheaper, and more environmentally friendly ways of doing things. It would mean, for the first time, living better electrically—or at least living better efficiently.

"Part of Ontario Hydro's challenge," wrote Marion Fraser, Hydro's energy management relations boss at the time, "is to promote the development of an 'energy efficiency industry' in Ontario in much the same way it created a nuclear industry 25 years ago."

The intricate political arrangements that governed Ontario Hydro, coupled with government's need to keep the lights on without big rate hikes, meant that any new energy-efficiency industry would demand the investment of significant political capital—even though the financial investment would be smaller than what it would take to keep Hydro's nukes going.

The energy-efficiency piece already had significant momentum. Hydro customers like Bob Lake's Peterborough Public Utilities Commission were finding ways of doing more with less power. Numerous voices, including some urgent ones in Ontario, were already pointing out that conventional energy economics was being turned on its head. It had to do with energy productivity. Between 1970 and 1998 Canada's Gross Domestic Product grew by 128 per cent, or twice as fast as the rate of primary energy consumption, which increased by 63 per cent. During the economic boom of the 1980s, when the Peterson Liberals enjoyed ample revenues, the electricity intensity of the provincial economy did not increase.

People who were examining these things noticed that in 1958 Ontario had put out just over $3 worth of goods and services for every kilowatt-hour of electricity consumed. By 1974 the rate had fallen to $2.35, which is where it stayed to the early 1990s. What this meant was that during the period in which Hydro's systems planners were being trained in the eternal verities of exponential growth in power demand, electricity consumption had grown faster than the economy. Then consumption levelled off, growing at the same rate as the economy. Rather than investing in nuclear and fossil-fuelled generating plants, it seemed to make far more economic sense to put money into energy efficiency: improved conservation standards for new buildings; retrofitting existing buildings with better insulation; more efficient, variable-drive electric motors in industry; compact fluorescent light bulbs; more efficient office equipment and home appliances.

Critics like Ralph Torrie had told the Porter Commission ten years before that it wasn't the buying of electricity that the owners of auto parts plants and home refrigerators were concerned about—it was running factory motors and keeping leftovers cool. Electricity is a means to an end rather than an end in itself. But at Ontario Hydro the significance of this idea had apparently escaped those who made the big decisions. Torrie observed:

> Well into the 1980s, the electric power industry misjudged what was happening in their marketplace on a monumental scale. . . . [They] did not see the change that was coming, or saw it coming but thought that electricity was a special case or that the decoupling would be a passing trend and then growth would return to its historical relationship to economic growth.

Perhaps it was naive to think that a company as large as Ontario Hydro, having achieved its size and power by selling more and more electricity, could learn to get into the business of selling less and less. A move in that direction would involve a major intellectual retooling: conceiving of energy supply not just as building central-station generating capacity but as something different; substituting more efficient uses of existing power for the need to generate more power. By the late 1980s this change in mindset (the cliché of the day was "paradigm shift") had started to make tentative inroads at Hydro. That was when Ontario Hydro published what turned out to be its last detailed plan. Conceding some ground to its insistent critics, the plan included a fat environmental analysis—albeit as an appendix—and bore a title hinting strongly of its even-handedness: "Providing the Balance of Power."

Commonly known as the Demand Supply Plan (DSP), it was an impressive document. Gigantism was clearly still a hallmark of the company's culture. The Plan weighed in at 2.8 kilograms. Each chapter opened with the soft hues of original artwork, each piece depicting pastoral vistas. Trees and foreground greenery dominated mine headframes, transmission corridors, belching smokestacks, and other signs of industrial activity. The section on demand management featured a painting of men busily insulating a new house out in the country. The days when marketing at Hydro meant selling baseboard heating and appliances so that customers could "Live Better Electrically" were, apparently, over. The Plan was unequivocal: "Hydro has no programs whose objective is to build load." Hydro planners were predicting that spending billions on conservation programs would reduce electricity demand by 5,400 megawatts by 2014, or about one-quarter of the anticipated increase.

But still those planners assumed that demand would continue to escalate. To meet this increase, Hydro came up with another major expansion plan, proposing ten new Candu reactors with a capacity of 7,100 MW (the equivalent of two Darlington-sized stations), in addition to thirty-two gas turbines to produce 4,300 MW. Hydro also assumed that the price of electricity would fall slightly over the course of the planning period, while the price of oil and gas would rise.

The prevailing belief still seemed to be that economic growth *required* the use of an increasing amount of electric power. Planners in the 1950s and 1960s had watched consumption grow by 7 per cent annually, in lockstep with impressive rates of GDP growth. By 1989 Hydro had scaled back its estimates of growth in demand, but was still counting on the houses, offices, and factories on the other side of the meter using 2.7 per cent more power every year until 2000. After that, the rate at which peak power demand would grow fell off slightly—though it would still keep growing. The guiding philosophy underpinning Hydro's DSP was, as it had always been, to meet the demand by going after economies of scale in generation.

The emphasis on demand management and energy efficiency seemed positive enough. But the DSP's authors pointed out: "The benefits created in any one community, however, may be lower than the direct impact of a major supply project. Demand management programs are unlikely to create significant local community impacts." Energy-efficiency programs are by definition decentralized and modest in scale, particularly when compared to traditional power-supply projects. To borrow a political metaphor of the time, such programs would shine a thousand points of light at a thousand places that called for improved electricity use. Central-station generators were akin to giant floodlights: much more noticeable. Getting supermarket chains and office landlords to replace wasteful fluorescent lighting systems with slimmer, shorter bulbs—as Hydro successfully did— had a lower profile than cutting the ribbon on a huge new generator.

Although the DSP was drawn up by people outside Hydro's engineering hierarchy and was comprehensive enough to become a Harvard

Business School case study, it was headed by Lorne McConnell, who had come to Hydro from Atomic Energy of Canada Ltd., which, after all, was charged with flogging Candu rectors overseas. Hydro engineers and nuclear boosters had little vested interest in programs that would curb the demand for power, and for more reactors. That was why, despite mounting pressure from both inside and outside the company, the final version of the plan included no scenario that envisaged lower growth in power demand; nor did it take into account the possibility of a low-growth future.

According to Rod Taylor, who worked on the DSP in the company's executive offices on the nineteenth floor, "the senior design and construction hierarchy at Hydro at the time" would not buy into that sort of alternative plan. Ontario Hydro was still characterized by institutional inertia. The company had grown into an engineering and construction operation whose principal goal was building big central-station power plants. Even when it came to demand-side management, there was a sense inside the company that the big consumers of power remained steel plants, mines, paper mills, and other large manufacturers. It had not yet filtered into the mindset of the planning hierarchy that retail operations with huge lighting, heating, refrigeration, and air conditioning needs had become crucial power-users. The engineers tended to see demand management as a frill made necessary by the changing politics of the day.

"It was very clear that the construction side of the house was not going to get approval for additional plants, more nukes or natural gas, unless they had a balanced plan," said Marion Fraser, who testified for six weeks at the environmental assessment hearings into the DSP. "Unless they at least appeared to be doing something on the demand side, they would not get approval on the supply side."

Parts of the utility were already starting to take energy efficiency seriously when the Plan was first tabled. Utilities in the United States had already begun to rethink the traditional view that their concerns ended at the customer's meter. Ontario Hydro was beginning to follow suit ten years after the Darlington protestors realized that "No Nukes!" had to be accompanied by positive Sun Day messages about renewable energy and conservation. It had taken Hydro a while, but in 1989 it launched its first province-wide demand-management programs, and there were signs that some within the organization were recognizing that conservation, rather than being a fashionable frill, was in itself a source of power that would allow the province to get much more out of the existing power system without spending hundreds of millions on new central-station generators. A "Savings by Design" program took aim at getting developers to build energy savings into new buildings and at starting an energy-efficient lighting program. Spending on DSM would be cost-effective. Unlike the Darlington plant and its accompanied interest-rate load, the program would lower—not raise—customers' electricity bills. It could be put in place more quickly than new power plants. It would be environmentally benign.

In the first three years of its modest DSM programming, Hydro successfully reduced customer load by 1,000 megawatts, which amounted to more power than any reactor in its increasingly unreliable fleet of Candus was able to put out. Within those same three years the company had thirty programs available to residential, commercial, and industrial customers, offering an array of information, energy audits, and financial incentives aimed at lowering the cost and risk of efficiency improvements. Moreover, the Ontario government was no longer in the hands of politicians with personal commitments to nuclear power and the old growth-at-all costs mentality. Green power and energy-efficiency advocates were positioned to question Hydro's plans directly at the upcoming Environmental Assessment hearings (1990–93) on the DSP.

The company was changing—but only sort of: Ontario Hydro was still a traditional electrical utility; it still held a monopoly on the supply and transmission of electricity to virtually every citizen as well as every business and public institution, small and large, in the province; and it was still intent on building more big generators.

Walt Patterson of London's Royal Institute for International Affairs has dissected matters electrical from plutonium and nuclear power to the role of electricity in the reconstruction of post-Stalinist Romania. Looking back at the gathering storm, Patterson noted in 2003:

> From the 1920s onwards the monopoly franchise, under some form of government regulation, became the basis of electricity systems around the world. It remained so until the late 1980s and the

advent of liberalization. . . . For generations, the monopoly franchise, with its captive customers, removed the risk associated with investing in larger and more long-term generation projects. In principle, such plants were expected to produce cheaper electricity. But even if they did not, captive customers would pay. The result was a steady increase in the unit size of individual generators and of the central stations housing them. From the 1950s through the 1980s, the scale-up was relentless.

The gigantism epitomized in Ontario by Darlington coincided with the arrival of what Patterson called the "liberalization" of British electricity—the privatization of state utilities and the introduction of a competitive power market that characterized Augusto Pinochet's and Margaret Thatcher's belligerent brand of capitalism. It also coincided with more fundamental changes in the way electricity would be generated as smaller, more decentralized turbines began to appear. Amory Lovins's book *Small Is Profitable* used a dinosaur metaphor: the great beasts that once dominated using their immense size and strength were replaced by smaller creatures better adapted to changing circumstances. The business press noticed the shift as early as 1980, when *Business Week* quoted a General Electric executive under the headline, "The Utilities Are Getting Small." No company had dined out on the gigawatt-sized building boom more than General Electric, but the company was sensitive to shifting trends. "Uncertainty over demand is the main reason for the appeal of small plants," said GE's Jerry Peterson. "If you're wrong with a big one, you're really wrong. If you're wrong with a small one you can just put up another."

A new generation of smaller, cleaner generators had arrived. They could be placed much closer to power-users, thus potentially transforming the grid that had emerged in the era of central-station power. The technology that most effectively captured the public imagination was the highly visible wind turbine. Denmark had already begun to develop wind power, which would give rise to a highly successful turbine manufacturing industry, but not without a protracted struggle between wind co-ops and the traditional monopolies of Danish electric power companies. Biomass—non-fossilized organic matter—could be either burned or gasified to power turbines. Photovoltaic panels on individual buildings could transform sunshine into electricity. Fuel-cell technology promised to produce power not through heat but through electrochemical reactions. Also promising—if not from a soft energy path perspective, then at least from the business viewpoint—were gas turbines that could be mass-produced and delivered to the customer's side of the meter for on-site generation. Such "aeroderivative" machines, adapted from mass-produced jet aircraft engines, would be small and efficient, offering co-generation to customers also wanting heat.

Add to these innovations the promise of improving energy efficiency through demand-side management, and it was becoming apparent that the soft energy path was a promising road. This was particularly true for those who took seriously the increasingly alarming evidence that burning fossil fuel was changing the global climate, with potentially catastrophic effects. It appeared that E.F. Schumacher, guru of environmental economics, had been right all along. Small really *was* beautiful. In 1974 Schumacher had written that modest operations were always going to prove less harmful to the environment than megaprojects because their individual effects would be small in comparison to nature's recuperative powers. Even when small undertakings commit environmental sins, Schumacher argued, they are "trifling in comparison with the devastation caused by gigantic groups motivated by greed, envy and the lust for power." Still, the intermediate technology pioneer remained optimistic about the capacity for human creativity in achieving what he called "an economics of permanence" instead of ever greater concentrations of economic power doing ever greater violence to the environment.

Ontario and its mega-utility had reached a watershed with respect to the politics of electricity. No political party was yet talking about Thatcher's private-power solution. The Liberals seemed to have broken with the steadfast Tory commitment to more and bigger nukes. They seemed open to major conservation initiatives, although Premier Peterson saw Hydro as a tool to be used for lofty reasons of state. Hydro not only ensured a steady supply of electricity at low prices; under both the Conservatives and the Liberals, the massive public investments in the nuclear program goosed the economy. According to Rod Taylor, Hydro's director of corporate strategic planning at the time, "The real corporate strategy of Ontario Hydro through the seventies and eighties was to be the biggest construction company in Canada, and what that meant was that governments of various stripes could use it as a countercyclical economical tool. Things get tough in the early eighties, you ramp up Darlington. Things start to go sweeter in the mid-eighties, you pull down Darlington."

Regarding it as his "fundamental obligation" to provide Ontario with cheap electricity, Peterson hoped to make a big addition to Hydro's supply side while helping to patch up the tattered national unity file at the time of the Meech Lake failure. He had already talked to Premier Robert Bourassa, who seemed amenable to a major transmission interconnect through Quebec that would allow the shipment of power from Labrador's lower Churchill River to Ontario. The Churchill had the greatest hydroelectric potential of any river on the continent, but Newfoundland had been badly burned on the hydroelectric development of the upper Churchill; the province was locked into an ill-advised contract to sell power at absurdly low rates to Quebec. Ontario would assist with hydro development on the lower Churchill. Peterson hoped to persuade Newfoundland Premier Brian Peckford to provide renewable power to Ontario and economic spinoffs to Newfoundland, build an east-west interconnect, and try to thaw frosty Quebec-Newfoundland relations.

Peterson later attributed his deal's failure to Peckford's "bitching, whining and complaining." He never got a chance to push the idea further because his government suffered a surprise defeat at the hands of the party that seemed to be very much open to a greener electricity policy for the province. As it turned out, the NDP would continue the Liberals' halting approach to creative power-planning. But, in the midst of crisis, the social democrats would also set the table for privatization.

five

The NDP Years: Clearing the Track for Privatization?

A society both expresses itself and structures itself by the technology it uses. When a decentralizing social trend runs up against a techno-logical trend that is inherently centralizing, or seen to be centralizing, feelings of alienation, loss of freedom, and abdication of responsibil-ity may be the result.
— Science Council of Canada, *Canada as a Conserver Society*, 1977

A HAZY MORNING SUN had not yet taken the chill out of the autumn air. Security guards waved cars away from King's College Circle at the University of Toronto, where Ontario's new government would be sworn in. Inside the beaux arts Convocation Hall a person or persons unknown had draped two huge banners over the railing of the top balcony so that the new ministers could not miss the reminder as they waited to take their oaths of office: "NO NEED FOR NUCLEAR" and "DON'T FORGET TEMAGAMI: EARTH FIRST."

The organizers had arranged for a chorus of schoolchildren and senior citizens to sing happy songs. Along with his few experienced ministers and some hacks in the press contingent, Premier Bob Rae noticed the faint laughter that rippled through the throng when the singers came to a verse in "Side By Side." "We ain't got a barrel of money, maybe we're ragged and funny . . ." Outside, where the cables from the TV trucks snaked through the fallen leaves, a platoon of maintenance workers scurried about with two-stroke engines strapped to their backs. As their gas-powered blowers roared and coughed acrid fumes, the leaves flew this way and that. Perhaps the New Democratic Party's campaign pledge to promote energy conservation would mean the return of the broom or the rake.

For some time the NDP had been organizing meetings with left-leaning greens to consider ways in which the party could take environmentalism more seriously. *Greening the Party, Greening the Province: A Vision for the Ontario NDP*, published six months before the election, recognized that social democrats and environmentalists were movements whose "critique of capitalism shares common themes." But tensions remained. The social democrats were committed to economic growth under "managed capitalism." An ever-larger pie could be shared more equally. The radical greens, with their concern about the Earth's limited carrying capacity, argued that the market economy had to be reigned in along with centralized mega-technologies.

"Our party stands at a crossroads," Bob Rae wrote in his introduction to *Greening the Party*. "Are we ready to become a green party, a party that puts as much emphasis on people's relationship with nature, and our obligation to future generations, as on the distribution of power and income, and the more traditional concerns of social democracy?"

On a less high-minded note, Treasurer Floyd Laughren used to joke that the first thing the NDP would do if elected would be to nationalize Ontario Hydro. His years on the opposition benches at Queen's Park had taught him a lot about Ontario politics, and one of them was that Ontario Hydro was a law unto itself. It was not long before the Rae government began trying to rein in the Crown corporation in an effort to make it more accountable and efficiency-conscious. The NDP government's first throne speech proclaimed a moratorium on new nuclear construction and mentioned ambitious energy conservation programs. The utility had previously earmarked $240 million for pre-engineering studies for new nuclear plants; now, somewhat sceptically, it dutifully reallocated that sum towards new conservation initiatives.

The new government's environmental commitments did not mean the cancellation of the nearly completed Darlington nuclear station. The ghosts in the machine at Darlington had all been exorcized, and cancelling the plant with over $10 billion already spent was a pill that Rae and company were not prepared to swallow—or that the new premier even wanted to try to swallow. "I was never anti-nuke," he later said. "At our party conventions people would get up and say, 'We have to shut her down.' My view always was that we have to take what's there and run it as efficiently and with as much responsibility as we can."

The new Darlington plant was the pride of Hydro's nuclear division, but as the literary critics tell us, overweening pride is an essential element in tragedy. This particular drama would feature the downfall of the old Ontario Hydro as Rae's pragmatic approach would ultimately result in the NDP taking the first steps towards the unthinkable—the privatization of the largest publicly owned utility in the country. But Ontario would not, as it turned out, start down the soft energy path and replace the behemoth with something inspired by *Greening the Party, Greening the Province.* Rae's accidental government, whose electoral platform would later be described by party insiders Chuck Rachlis and David Wolfe as "little more than an election ploy," gave little serious thought to greening the power system and was wide open when another road presented itself.

"It was such a surprise that we got elected that we had not set our own priorities," said Bud Wildman, who held the energy portfolio and was one of the NDP's most effective cabinet ministers. "We were as surprised as anyone else."

The NDP's approach to economics was not far from the mainstream of Canadian politics as practised by the Ontario Conservatives under premiers like Bill Davis. The social democrats sought to push for economic management with a human face by injecting a concern for equity into government policy. They sought to promote and defend more equal taxation, strong protection for the rights of workers, and an active, albeit pragmatic, government role in regulating the economy. Once in power, though, the party staggered from crisis to crisis. Its members launched consultations without a clear plan of the action on what would follow and backtracked on the central elements of their campaign platform. In short, they got spooked.

It was a scary time to be a social democrat. In 1990 the Ontario economy entered its worst recession since the 1930s. Provincial unemployment levels rose from 5 per cent in 1989 to over 11 per cent by 1993, while government tax revenues dropped and social expenditures rose. The federal government reduced its transfer payments to the province for health care and education, sending the provincial deficit even higher and creating fears of a "debt wall" within the NDP cabinet. Electricity prices skyrocketed as soon as power started to flow from Darlington.

Beyond the economic downturn loomed a global economy that seemed to render traditional social-democratic strategies useless. After proudly proclaiming in their first budget that they would introduce Keynesian-style spending programs—which meant they would fight the recession rather than the deficit—the Rae government soon began to rethink its approach in the face of concerted attacks by business lobbies and the increasing acceptance of neo-liberal arguments against deficits within the media, the public at large, and even the party. The government that had sought to manage markets as a way of protecting the citizenry was now prepared to be managed by the market.

New Democrats were well versed in energy policy. Throughout their long opposition years, they had participated in apparently endless royal commissions and select committee investigations into Hydro. *Greening the Party, Greening the Province* contained a strong commitment to energy conservation, proposing the reduction of the province's energy intensity by 3.5 per cent per year over the following twenty years. The government's moratorium on nuclear construction and the appointment of Marc Eliesen as Hydro president and chair were further evidence of its resolve. Eliesen was a career NDP bureaucrat with considerable utility experience, including a stint as chair of Manitoba Hydro.

But there were early signs of disquiet. The natural choice for the energy portfolio and Hydro overseer was former energy critic Brian Charlton, but he blotted his copybook by musing publicly about conservation measures before the cabinet-making process was finished. In a petulant moment, Rae decided to discipline Charlton by keeping him out of the cabinet for a time. Rae's choice for energy minister was environmental advocate Jenny Carter, an untested political novice. Carter got a rough introduction to politics when she mused—again publicly—that her husband Cyril (a physics professor at Peterborough's

Trent University) would make a better energy minister than she. Carter didn't last long in the portfolio, and Eliesen left for B.C. Hydro in 1992.

The government soon passed a bill amending the Power Corporation Act to increase the accountability of the stubbornly independent Crown corporation and encourage energy conservation initiatives. Although the conservation thrust left Hydro's nuclear boosters more than a little perplexed as they tried to adjust to the notion that they should help their customers buy less power, the most politically controversial change was one empowering the minister to issue policy directives to the Hydro board of directors. The Board would then have to implement those directives. To make the whole process more transparent, the directives were published in the *Ontario Gazette*, the province's official notices publication.

The NDP increased the size of the Hydro board of directors from seventeen to twenty-one, making room for Aboriginal, environmentalist, and labour representation and extending representation beyond the businessmen traditionally appointed to the body. The NDP's initial appointees to the board also included, for the first time, a strong representation of women. The deputy minister of energy was appointed to the board as a non-voting member to provide a "window" for the government into Hydro activities. Hydro historian Neil Freeman notes the irony of this amendment, first proposed in 1972 so that the deputy minister could inform Hydro of the government's plans. It was now finally being instituted—but in an effort to make sure that the government was informed of Hydro's plans.

One of the appointees was David Brooks, the founding director of Ottawa's Office of Energy Conservation. Brooks agreed to join the board after Eliesen assured him that it would be okay to discuss non-confidential matters with an informal reference group of fellow environmentalists. At first Brooks found it hard to comprehend numbers that were beyond anything he had seen while pushing the soft energy path in Ottawa during the 1970s. He was dealing not with millions but with hundreds of millions of dollars. The company's huge entrenched bureaucracy was "efficient and competent" on its own terms, he found. "You also had a generation of engineers who genuinely thought nuclear power was going to save the world, providing electricity too cheap to meter."

Brooks, who saw himself as a moderate among environmentalists, was still unprepared for what he observed when he got into the belly of this huge, self-assured beast. Darlington had just opened after years of billion-dollar delays. "The problems with Darlington were seen as glitches rather than fundamental conceptual flaws. And the depths of the problems at Bruce and Pickering were not yet apparent." (In 1997 seven reactors at these plants were shut down due to safety concerns.)

The utility's governance did not lend itself to scrutiny by its own board, let alone the public at large. Even for Brooks, a resource economist and ideas man whose first two degrees (from MIT and CalTech) were in the natural sciences, the mysteries of nuclear power remained difficult to fathom, particularly when framed by the staggering amounts of capital at stake and explained by men committed to the controversial technology. "It's very easy for an entrenched bureaucracy to put things over on a board, especially within an area as financially and physically complex as a nuclear reactor and when there are strange things going on in the chemistry." Brooks found himself wondering if the senior staff were even aware of the depth of the problem. "Were they groping as we were? Certainly I'm now convinced that the confidence they displayed in their solutions really meant that they were lying to us."

Along with making the board more representative, the NDP also implemented the previous government's plans to expand power generation by giving independent power producers easier access to the grid. Despite Adam Beck's best efforts, Ontario's electrical system had always had some private power generation. Now the government was taking a cue from U.S. regulators and telling the Hydro monopoly to buy power from non-utility generators. The idea was to provide a counterweight to Hydro's influence, introducing a hint of competition in the hopes of keeping Hydro honest. NUGS, particularly the ones that were working to install gas co-generation and renewables, were seen as one way of steering Hydro away from its obsession with mega-projects. Brian Charlton, who eventually did make it into the energy minister's chair, took pride in how the government was the first to bring private-sector generators into the system.

This was not, however, a holus-bolus rush to a competitive market. The NDP sought to maintain a dominant role for the public sector, particularly in determining new sources of supply. Charlton recognized that a straight market system unaccompanied by government guidance would discriminate against green power. "On a stand-alone basis a small hydraulic electric generating system is not competitive with some very dirty fossil fuel alternative unless you're prepared in some kind of a regulatory system to award the environment points that the private marketplace won't recognize." The NDP basically retained a commitment to a predominantly publicly owned—and more highly regulated—system providing power at cost. But the government sought to broaden what was meant by cost to include social and environmental criteria. Its initial approach to electricity policy was to change Hydro's mandate from energy supply to energy services while keeping both the economic and environmental costs down. By continuing with the environmental assessment of Ontario Hydro's twenty-five-year Demand Supply Plan, the Rae government was providing environmentalists and other interested parties with the opportunity—and millions of dollars in funding—to hold Hydro to account in a set of complex hearings; the greens used the opportunity to launch an unprecedented assault on the rationale underlying the utility's bigger-is-better, growth-is-forever mindset.

The Hydro establishment did not take all of the new government's changes lying down. As soon as Rae appointed Marc Eliesen as chairman, the board countered by naming Al Holt its chief executive officer. Rae later said he had nothing against Holt, a prominent member of the utility's old guard, but: "It was a runaway board, patting the government on the head." From Rae's point of view, the board was more or less saying, "Don't worry sonny, we know how to run this thing." Rae regarded the board's behaviour "as almost a kind of a coup" and told himself, "This ain't going to happen on my watch."

The Hydro brass had an angry premier on their hands. Historically the relationship between the government and Hydro had always been based on the relationship between the premier and the Hydro chair, and matters proved to be no different with the Rae government. Eliesen was never well-loved by the Hydro board members, who believed he had been imposed on them, and he had the added misfortune of leading the company at the time of the post-Darlington rate hikes. He found himself urging the premier not to hold back on price increases, arguing that taking heat for 14 per cent was the same as taking it for 10. No matter how sound the advice, or how inevitable the increases, it was not something likely to endear him to the person taking the heat. Eliesen found himself and the premier at odds. Then the Hydro chair found himself being attacked for his $400,000 salary—even though it was $140,000 less than his predecessor had received. For Eliesen it was a case of the government being spooked by the deficit. Rae thought that Eliesen could not adapt to the new economy and the role that Hydro would play within it: "Eliesen had a classical view of Hydro as a Crown corporation, as an empire, and he was not going to rock that part of the boat. In fact, he was not going to rock many parts of the boat at all."

When Eliesen soon left to become the CEO of B.C. Hydro, Rae found himself looking for someone new. His goal was to find someone "who would begin to create a different culture." He was facing, on one side, "negative feedback from businesses about Hydro's attitude," and, on the other, attacks on Hydro's lack of environmental concern.

Enter globetrotting businessman-environmentalist Maurice Strong. An old family friend of the Raes, Strong was recommended for the top Hydro job by party guru Stephen Lewis. For Rae it was "the smartest single appointment I made as premier." In Strong, Rae found a chairman and CEO who would not only rock the boat, but also start to scuttle it.

According to Hydro insider Rod Taylor, Strong came in and did what was necessary: "Turn the corporate strategy, unstated as it was, from being Canada's biggest construction company into what he wanted it to be—an efficient operating company." Strong also opened up the rate structure and introduced the possibility of privatization.

Strong was a man of apparent contradiction. A high-school dropout with forty-one honorary doctorates, he was a successful oil entrepreneur whom United Nations Secretary General Kofi Annan described as "one of the world's leading environmentalists." Originally a poor boy from the Prairies, he ended up presiding over the single largest gathering of heads of state in world history, at the 1992 Rio Earth Summit. Pilloried by the left for his support of corporate-driven environmental regulation, he also figured prominently in the paranoid fantasies of the far right in the United States as a shadowy figure attempting to impose world government at the behest of a power-hungry UN. The ultimate insider whose Rolodex boasted a who's who of global geopolitics and industry, he established his own commune and would go on to be a close advisor to Prime Minister Paul Martin.

Strong's career followed a wildly improbably path. By age thirty he was CEO of Montreal-based Power Corporation and hobnobbing with Canada's business and political elite. He hired the ambitious young Martin as his executive assistant in 1962. Strong then did one of his sideways leaps, joining the federal civil service, where he helped to establish the Canadian International Development Agency. After serving as secretary general for the first Earth Summit in 1972, he moved back and forth between the UN bureaucracy and the corporate boardroom.

Shortly after taking the helm at Hydro, Strong pronounced it a "corporation in crisis." At the heart of this crisis was the Darlington nuclear station. Hydro's rate rules were clear: the cost of a new facility could not be included in electricity bills until that plant was actually producing energy. In the ten years in which Darlington was under construction, its mounting price tag was simply racking up interest payments with no impact on the price of electricity—out of sight and out of mind for everyone except Hydro planners and a handful of civil servants. But once Darlington came on line—which happened to be over the first three years of the NDP government— electricity rates skyrocketed to pay for power that wasn't needed.

Consumers and businesses began to scream. The price of power rose by 31 per cent between 1991 and 1993, and Ontario went from a low-cost power jurisdiction to one that was above the North American mean. At the same time the demand for electricity flattened or fell. Not only had Hydro's forecasters overestimated the growth in demand, but Darlington's power began feeding into the grid just as demand was being reduced by a deep recession. Hydro—and by extension the government—was stuck with a whopping 3,500-megawatts surplus and rapidly rising rates in a time of recession and low inflation.

To make matters worse, many large industrial customers were considering generating their own power with natural gas, which was suddenly cost-effective given Ontario's higher, nuclear-boosted rates. Add in a huge debt burden that was the product of the long years of expansion, and you had a recipe for a "debt-death spiral," which went something like this: in response to higher prices, big Hydro customers would produce their own power using the new natural-gas co-generation plants (a gas glut was pushing prices down at the time); this new activity resulted in a smaller revenue base from which to repay the huge nuclear debt; rates were forced even higher; more customers left, accelerating the cycle. The end result was a provincial government, and by extension the taxpayers, stuck paying for the white-elephant nuclear plants that were producing power no one wanted to buy. Cries of "I told you so" from the people who had been warning that the nukes were a pricey folly were met with stony silence.

Within the government, two camps emerged. A "bail-it-out" group wanted Hydro kept as a government monopoly and called for the province to take over the bulk of the utility's $26-billion debt. Ontario was on the hook anyway; it provided the debt guarantees. A "cut-it-loose" camp, led by Deputy Health Minister Michael Decter and then-Deputy Minister of Finance Eleanor Clitheroe, called for Hydro's monopoly to be abolished. The government should sell shares in Hydro and promote a competitive market. Although the NDP had traditionally supported a larger role for private power producers (the non-utility generators) within a publicly controlled system, what was being talked about here went beyond adding a few private players. It would mean transforming the system from being publicly owned, with private producers on the edges, to a predominantly private system—like the one being implemented in the United Kingdom.

Within cabinet, both Energy Minister Bud Wildman and Finance Minister Floyd Laughren were opposed to privatization, and Laughren told Strong that there was no way privatization would ever make it through cabinet. The premier was at least considering it, primarily because Strong was whispering sweet pro-privatization nothings in his ear. Mostly, however, the premier wanted Hydro "out of the headlines." He informed Strong that there would be no privatization of Hydro in the first NDP mandate.

Strong moved quickly on three fronts. He froze hydro rates, laid off thousands of Hydro staff to cut labour costs, and rolled back spending on the conservation programs. The idea was to have a stable rate structure. "We very much needed to send a message to the organization that life was not going to go on the way it had gone on before," Rae said later. "They needed to become more supple and flexible and demonstrate a greater willingness to come to terms with this new economy." The status quo, as they invariably say, was not an option.

The premier's instructions to Strong reflected the atmosphere of the time: "Look, we have to recognize that there's a trend going on in every part of the economy and it would be pretty hard if it wasn't going on in Hydro. It's going on inside government—we're doing the social contract."

The social contract—which forced civil servants to take unpaid holidays—was mild compared to the bloodbath at Ontario Hydro, where Strong slashed over 30 per cent of the workforce in 1993 and 1994. The layoffs were concentrated in the nuclear construction department, which was largely redundant because Darlington was finally complete and the power surplus had thrown future expansion plans out the window.

The rate freeze and layoffs addressed the economic and political issues. Rae's political calculus was correct. As soon as prices stabilized, customers stopped complaining and the battered provincial economy maintained a key competitive advantage. The Opposition inevitably turned its critical questions elsewhere, and Hydro disappeared from the political radar screen for nearly ten years. Yet the environmental issues did not go away. Hydro's coal-fired plants continued to spew out 15 million tonnes of greenhouse gases annually. Its nuclear reactors—Candu plants seem to last about as long as an average plough horse, without the dependability—were starting to break down but continued to pile up hazardous waste.

As the winner of multiple environmental awards, Strong seemed at first glance to be just the man to put an end to Hydro's brown decades. All the greens knew that "Chairman Mo," as he was known to Hydro insiders, was still basking in the reflected glory of the United Nations' 1992 Earth Summit, where he played a key role in promoting and shaping one of the great buzzwords of the day: sustainable development. The concept had been popularized by the 1987 UN-sponsored World Commission on Environment and Development, which defined it as "development that meets the needs of the present without compromising the ability of future generations to meet their own needs."

If that seems vague, it was meant to be. The ambiguity inherent in the term, and the resultant political flexibility, meant that sustainable development could accommodate almost anyone or any direction. In the words of one jaundiced observer: "Sustainable development is a 'metafix' that will unite everybody from the profit-minded industrialist and risk minimising subsistence farmer to the equity seeking social worker, the pollution-concerned or wildlife loving First Worlder, the growth maximising policy maker, the goal orientated bureaucrat, and, therefore, the vote counting politician." With its promise of balancing economic and environmental concerns, the fuzzy notion quickly became popular with corporations.

Strong had organized corporate participation debates surrounding sustainable development before the Rio meeting. To ensure that the voice of big business was heard at the Summit, he helped pull together executives from Du Pont, Dow Chemical, Ciba Geigy, ALCOA, and a host of other companies with dubious pollution records, getting them all to join the Business Council on Sustainable Development. The enterprise was assisted by the good offices of Burson-Martseller, the public relations outfit hired to burnish the reputation of Argentina's infamous military junta. The idea of the Business Council was to recapture the environmental initiative from activists and regulators, repositioning big business as environmental leaders rather than as villains. Not surprisingly, the multinational corporations emphasized the importance of interpersonal dialogue, minimizing the role of regulation. Where change is required, it is to reduce the role of governments to create space for the new environmental champions and to redesign environmental policies along more market-driven lines.

At the UN's Rio conference, Secretary General Strong warned official delegates that any global environmental treaty would have to be "GATT-legal." (GATT would soon morph into the World Trade Organization.) "Strong's meaning was clear," green journalist Tom Athanasiou observed in the wake of the Summit. "The Earth Summit treaties would not be allowed to contradict the doctrine of free trade as codified in secretly negotiated trade agreements interpreted by an unelected cabal of unknown, Geneva-based bureaucrats." Strong, the consummate UN insider, simply assumed what Athanasiou made clear: "It was GATT, not the Summit, that embodied the emerging future."

That future, soon to emerge with stark clarity in Ontario, privileged the unfettered free market as the saviour of the environment. The new laissez-faire assumed that unleashing capitalism would encourage innovation and generate new, super-efficient technologies to save the environment without anyone having to reduce consumption. The prophets of free-market environmentalism were urging an end to cumbersome regulations that actually tell companies what to do. Instead they would replace those regulations with economic incentives modelled on the market, such as emissions trading, in which companies buy and sell the right to pollute. Natural resources should be privatized and assigned a dollar value. Environmental responsibility, in turn, should be shifted away from people acting as citizens through democratic means like governments or public hearings. Instead it should go to consumers who will duly influence corporate behaviour by buying green products.

When Strong came to Ontario Hydro, he brought along this sort of free-market thinking. He had always been a "big ideas" man more interested in structural reform than day-to-day management. Many insiders found Strong's arrival a breath of fresh air. "It was an incredible time," said Allan Kupcis. "Strong had both the agenda and the mandate to stop the financial bleeding. His job was to stop industrial customers from raising hell with Bob Rae."

Other so-called stakeholders qualified as second-rate. Strong cancelled Ontario Energy Board hearings and the Demand Supply Plan Environmental Assessment, dismissing this form of public participation as "apparently endless," beset by "interventions by environmentalists and other interest groups." At the same time he eagerly launched into a corporate visioning process designed to transform Ontario Hydro into "a leader in energy efficiency and sustainable development." The resulting discussion paper on the corporation's future—"Hydro 21: Options for Ontario Hydro"—was chockablock with buzzwords about eco-efficiency and competitiveness as the saviour of the environment. Strong set forth his reasons for rewriting Hydro's mandate as opposed to trying to make the utility's conservation programs work:

> The changes we're making provide a unique opportunity to integrate sustainable development and eco-efficiency practices into our whole organization. We must get out of the habit of thinking of them as "add-ons." To be effective, they must be pervasive. They will, in fact, contribute significantly to the goals of our change process.

The "add-ons" that had to go as part of this revisioning process included investing money in energy conservation. Perhaps only Maurice Strong could have gotten away with slashing the utility's environmental programs in the name of sustainable development. The NDP had come to power with a commitment to take the promotion of energy efficiency seriously, and it had changed the Power Corporation Act to broaden the definition of conservation, permitting Hydro to promote a broad spectrum of energy conservation programs and not just those involving the use of electrical energy. The government had also paved the way for a more activist Hydro, removing a clause that had prohibited loans, incentives, or technical assistance for conversion of electric space heating to different fuels. That change may seem trivial, but the idea that Ontario Hydro would help people switch from electricity to a cheaper, cleaner fuel was revolutionary. A 1992 study revealed that in the previous fifteen years, Ontario consumers with electric heat (disproportionately low-income households) had spent almost $4 billion more than they would have spent if they had heated their homes with oil or natural gas. If a fifty-fifty mix of natural gas and oil heating had been substituted for the largely coal-fired electric heating between 1975 and 1991, about 75 million tonnes of global-warming-causing carbon dioxide emissions would have been prevented. Roughly 1.3 million tonnes of gases that caused smog and acid rain would have not found their way into lungs and lakes. The NDP's changes also allowed local utilities to treat the costs of conservation programs as a capital expenditure, putting conservation projects on a more equal footing with the building of new supply—the projects would not have to pay for themselves within a single year.

The government had initially cranked up the energy conservation programs begun under the Liberals, launching a Green Communities program so that grassroots organizations could help deliver the "decentralized megaproject" that is energy efficiency. Environmental activists started to trade their placards in for caulking guns and batts of insulation. It seemed to be time to stop talking about the Conserver Society and start making it a reality.

Energy Minister Wildman noticed that Hydro's leadership still had trouble embracing the twin concepts of encouraging customers to use less power and doing things in a small way. "For Ontario Hydro, everything is a megaproject," Wildman said. Marion Fraser, an old Hydro hand who ran Hydro's conservation programs, recalled the internal dynamic at budget-making time. "It was probably easier to get a billion than a million, especially on the supply side."

Yet even with these imperfections, the demand-reducing programs were having an impact, much to the surprise of the traditional supply-siders. By 1991 Ontario Hydro was investing $179 million in conservation and energy-efficiency programs. This money was helping two hundred communities upgrade their streetlights to more efficient bulbs. Along with a range of other initiatives, the lightbulbs would save Hydro's customers $28 million on their energy bills annually and reduce demand by 250 megawatts—enough power to meet the daily needs of a small city. Long after Strong had left the scene, energy-efficiency policy wonks were still talking fondly of the small town of Espanola, where 86 per cent of citizens and businesses had become involved in an Ontario Hydro program. Electricity use in the town was reduced by 17 per cent.

By 1992 the cumulative savings from Hydro's conservation programs since 1988 were reducing customer bills by $260 million and cutting demand by 800 megawatts, or enough power to run the nation's capital. There were five times as many energy service companies as there had been five years previously, and there was no sign of the energy-efficiency push running out of steam. Hydro was forced to issue an update to its Demand Supply Plan. The update nearly doubled the forecast contribution from energy efficiency and acknowledged that, based on "the success to date in an ambitious program to help customers use electricity more efficiently," the push to efficiency could save more energy than new nuclear or coal plants would produce by 2014.

In October 1992 Ontario Hydro hosted an international conference on energy efficiency and demand-side management to showcase its initial successes. It was the first international conference on the subject and, according to Fraser's nostalgia-tinged recollection, "It was like one brief shining moment where everything came together. People came from all over the world to look at what we were doing and how we were doing it."

It was, as it turned out, a fleeting moment. The "thousand points of light" decentralized approach to conservation, efficiency, and generation from renewables that was the hallmark of the soft path would be doused by one of the world's best known environmentalists. Maurice Strong was at the October conference, getting to know the company he was preparing to take over. After assessing Hydro's debt situation and placing that against spending money to reduce demand, he decided to eliminate what he called "subsidies" for energy efficiency. (The vocabulary is noteworthy. Free-market fundamentalists predictably refer to any state spending that they disagree with as a subsidy.)

Hydro's spending on energy efficiency was replaced by the rhetoric of eco-efficiency and sustainable development. By 1994 spending on energy-efficiency programs had been cut back to $94 million, half of what it was in 1991. The programs were reframed as something that Ontario Hydro had gotten into when the recession had made it "hard to find the necessary capital to invest in energy efficiency" and when the efficiency marketplace was "co-ordinated poorly, if at all." Now that the energy retrofit business was "a vibrant, $1 billion dollar business in Canada, mostly in Ontario," Hydro stated, "broad-spectrum financial incentives from Ontario Hydro are no longer required to promote awareness of energy efficiency." Rather than helping its customers use less of its product, Hydro would work at improving its own internal energy efficiency so that it would have more power to sell. Overall, Hydro's spending on environmental programs (which included both pollution-control equipment and energy conservation programs) went from 2.5 per cent of total expenditures in 1988 under the Liberals to 9 per cent in 1991 (before Strong took over) before going back down to 5.5 per cent of total spending in 1994.

Strong hired John Fox, who had managed highly successful energy conservation programs for a California utility. Fox was handed the "difficult task of overhauling and strengthening" the conservation program and phasing out direct subsidies that Hydro had been offering. Difficult indeed: California was spending hundreds of millions of dollars annually on energy conservation programs because they were cheaper than building new generating stations, but Fox was now expected to produce results without resources. "Maurice came in and really significantly curtailed a lot of construction activity and a lot of the operating budgets," Fox said. "The cost-effectiveness of some energy-efficiency programs did not hold up in that environment and therefore were significantly scaled back. The company maintained a commitment to efficiency, and part of my job was to do it better, smarter, and cheaper."

Not all was lost. Fox, a Canadian who served on the board of Amory Lovins's Rocky Mountain Institute, built links with trade associations and environmentalists, spawning the Canadian Energy Efficiency Alliance. The government-led programs such as the Green Communities Association continued and even flourished because they were able to expand their base of support to include local governments and municipal utilities. Rae's government amended the

Energy Efficiency Act to provide modest improvements in efficiency regulations for some consumer and commercial products, although to avoid angering the powerful Homebuilders Association the administration shied away from introducing serious energy-efficiency measures into the Ontario Building Code. The NDP continued Ian Scott's efforts to make Hydro's operations more transparent, in the hopes that this would prevent a return to the days when it could run its own show in closed-door collaboration with its masters across the street at Queen's Park.

Yet during the NDP period Hydro's political masters did not order Hydro to continue ramping up its conservation programs. "The recession and its impact on government spending, and the glut of power on the system, moved to the back burner the pressure for major environmental efficiency initiatives," said former Energy Minister Brian Charlton.

There is an irony here. Conservation in Ontario was suffering from a kind of reverse Keynesianism. When the economy boomed, there was little interest in conservation or efficiency because energy was too important to become a bottleneck holding back industrial growth; new power plants were commissioned to ensure that demand could be met. During a recession, on the other hand, you could not afford to try to cut back consumption further because you needed the revenue from the sale of electricity to cover the capital costs incurred in building new plants during the boom.

That is why environmentalists had pushed for an environmental assessment of Hydro's Demand Supply Plan. They wanted to point out the irrationality of the short-term, supply-side logic that had driven Hydro's endless expansion. The NDP's announcement of a nuclear moratorium in 1990 had lent an aura of unreality to the Demand Supply hearings, which were based on a plan to build new nukes; as a result there was little fuss when Strong put an end to the hearings three years later, complaining about bothersome interest groups. To the environmentalists who were developing an alternative energy plan, it was a pyrrhic victory. They wouldn't get the chance to fully debunk Hydro's plan and properly present their alternative, but the powers-that-be were at least admitting that they had been wrong on the need for ten new nuclear reactors and thirty-two turbines fired by fossil fuel.

Another game was also afoot. Having dispensed with public participation on the environmental assessment front, Strong discovered that he didn't even have to go through the motions of an Ontario Energy Board hearing. After all, during inquiries by the public regulator consumer advocates and environmentalists could ask questions about Hydro activities when the utility presented applications for rate increases. Chairman Mo said that he was "delighted" not to have to bother because the hearing would have delayed his restructuring plans: "As a result, paradoxically, the most extensive program of transformation in Ontario Hydro's history was carried out more rapidly than anyone thought possible, without prior review by the [Ontario Energy] Board, something that I don't believe could have happened in any other jurisdiction."

The restructuring was not limited to downsizing the nuclear construction division. Strong saw open markets and competition in electricity as inevitable, and there were now insiders at Hydro who were of the same opinion. One was former deputy minister of finance and now Hydro vice-president Eleanor Clitheroe, an ambitious former banker whose career at Treasury prospered during the NDP's female-friendly regime. She was sponsoring academic seminars and a text on *Ontario Hydro at the Millennium: Has Monopoly's Moment Passed?* And Strong was shuffling the deck of Hydro's internal structure to prepare it for privatization. Hydro was reorganized into semi-autonomous business units, with transmission and distribution separated from the generation side of the business. Strong created an electricity exchange and developed transfer pricing among Hydro's business units to duplicate competitive conditions in an internal setting. He even embarked on an hourly spot-market experiment to develop experience with competition among other Ontario generators, and prepare for what he saw as the inevitable opening up of Ontario's electricity market.

In a certain sense, the free-market advocates were aided by the success of environmentalists in laying bare the irrationality of the hard energy path. The North American electricity status quo had been given a good shake in the 1970s as three factors converged.

First, electricity costs increased due to the OPEC oil price hikes and the unexpectedly high capital costs of building nuclear plants. Second, technological progress in big generating equipment had stalled, casting doubt on the "bigger is better" strategy for building plants and boosting demand, which was premised on the assumption that continually increasing consumption of electricity would result in continually declining per kilowatt costs. In short, electricity was not going to become too cheap to meter, so it might make sense to start paying attention to what was happening on the user's side of the meter. Thirdly, the energy crisis had granted heightened visibility and credibility to the environmental movement and its questioning of the ideology of growth.

Initially, this shakeup led to thinking about a soft energy path and the gradual introduction of conservation programs into what had been supply-side-driven utilities. In the United States, it had also led to the 1978 Public Utility Regulatory Policies Act, which created a space for independent power-producers to bring power onto the grid, in what was billed as an effort to motivate innovation in small-scale technologies outside the realm of the traditional utility system. The mammoth Tennessee Valley Authority cancelled the construction of nuclear plants that were near completion and launched conservation programs instead. In Ontario the shakeup led to the grudging introduction of conservation programs on the ground and initial moves to integrate them into system planning at Ontario Hydro. There were also the new openings for non-utility generators and for municipal utilities to develop local sources—along the lines of Bob Lake's bid to bring small-scale hydro on-line in Peterborough.

In the end Strong's legacy as boss of Ontario Hydro was decidedly mixed. He brought a distaste for grassroots politics and a penchant for elite consensus-building. In three years he managed to dismantle both the nuclear construction division and the energy-efficiency programs. While extolling the virtues of transparency and the value of working with civil society groups, he cancelled the hearings that would have allowed public input into Hydro affairs. After agreeing to run a publicly owned utility on behalf of a social-democratic government, he pushed privatization behind the scenes and quietly reor-

ganized Hydro in a way that anticipated the heady business of selling it off to interested buyers.

By the time that Maurice Strong scuttled Ontario's conservation programs, the front men for big business had glommed onto the rising
wave of privatization talk. In any case, they thought, conservation
programs were overly interventionist. Government should never have
considered such subsidies in the first place. In fact, it should get out
of the business of running electricity systems. Who needs energy
planning—soft path or hard—when the invisible hand of the market
can provide cheap electricity better than any planner? Their desired
approach was often framed as a consequence of technological development—combined-cycle natural gas generators had rendered monopol
ies obsolete, they argued—or as part of a desire for "choice" on the
part of consumers.

Even though Rae was making it clear that an NDP government
would not be privatizing Hydro any time soon, Strong reached out to
bankers, private power companies, and regulators from privatizing
jurisdictions. Rather than looking to the energy-efficiency industry
that had emerged in the 1980s, he called on William Farlinger.

Farlinger was chairman and CEO of Ernst and Young, Ontario
Hydro's external auditor and the Canadian arm of the global consulting and accounting firm. He was also mentor to Mike Harris, the
leader of what was then the third party at Queen's Park. Harris was
attractive to Farlinger because the Member from North Bay—who
had arrived at Queen's Park with some crude conservative inclina-
tions of his own—was an affable golfer who had been recruited to
the neo-conservative cause by Tom Long, a brash right-winger who
represented the Thatcherite wing of the Tory party. While Long and a
group of up-and-coming whiz kids provided Harris with a set of simple ideas to sell to the electorate, Farlinger provided the Bay Street
contacts that funded the future premier's leadership bid and subsequent campaigns.

Attracted by Farlinger's Tory ties, Strong picked the head of Ernst
and Young to co-chair a Hydro Financial Restructuring Committee

that reported in June 1994. Six months later Strong hired Farlinger
to prepare a report on the future of Hydro. "Strong knew that an election was coming," Kupcis said. "Maurice knew that there was a good
chance that Bob Rae wouldn't get re-elected and he wanted to be
ready to push through reform regardless of who won the election."

Labelled "Privileged and Confidential," *Ontario Hydro and the
Electric Power Industry: Vision for a Competitive Industry* was submitted
on June 22, 1995, two weeks after Harris's election. It was a remarkable document for the sweeping changes it recommended, the lack
of empirical data to back up its claims of cost-savings, and the narrow range of stakeholders whom Farlinger consulted in its preparation. The report identified ten groups as having a significant stake in
the future of the province's electricity industry, including customers,
municipal electrical utilities, municipal governments, the provincial
government, unions, and environmental groups. Yet Farlinger and
his two co-authors (a utility specialist for Ernst and Young and a
former investment banker) restricted their consultations to private
power companies like Enron; banks and consulting firms such as
Goldman Sachs, Wood Gundy, Citibank, and J.P. Morgan; Moody's
investment rating service; pro-privatization U.S. academics; regulators from privatizing jurisdictions in the United States and United
Kingdom; provincial utilities; and Ontario Hydro insiders such as
Strong and executive vice-presidents Clitheroe and Dave Goulding.
(Goulding would go on to run Ontario's Independent Electricity
Market Operator.) Of the seventy-four people they interviewed, none
represented consumers, municipalities, local utilities, unions, or
environmental groups.

Given the ideological predilection of the writers and the advice
they sought, the conclusion of the panel was hardly surprising.
"Ontario Hydro must prepare itself to compete fully in an open market environment for generation in Ontario," they proclaimed. "This
view is shared by the senior officers of Ontario Hydro, who believe
that competition will come within 3 to 5 years."

Declaring the status quo unacceptable, the report went on to
describe the "inevitability of increased competition" and to argue
that competition would inevitably force electricity prices down.
While acknowledging that Ontario's electricity rates were lower than

the rates in competing U.S. jurisdictions, it claimed that the deregulation then underway in the U.S. Northeast would result in lower rates that would undercut Ontario and devastate its industrial base unless the government acted quickly to open the grid to market forces. The writers dealt with environmental protection in a single dismissive paragraph, and batted aside the financial arguments in favour of public ownership (lower public borrowing costs, no need to pay taxes or dividends) with a stunningly simplistic statement of faith in the magic of the marketplace:

> The arguments around privatization of Ontario Hydro are often aimed at trying to determine the net numerical cost advantage of public ownership which is typically related to freedom from income taxes and lower costs of government capital. However, the most compelling arguments are in fact the obverse of this approach. That is, why should any competitive service remain in government hands? Outside of electric power, there are few examples of competitive services that are successfully offered by government-owned enterprises. And those that have been in the past (e.g. Air Canada, CN Rail) have been or are being privatized. The reasons for this are simple. In a competitive environment, the marketplace works.

The marketplace can work in mysterious ways. Monopoly's moment did indeed appear to be passing. The Age of Enron was at hand.

six

South of the Border: California, Enron, and the Yankee Alternative

Speaker number one: "We decided that the prices were too low on the daily market, so we shut down everything except the Ormond plant."
Speaker number two: "Excellent. Excellent. Looks like the price popped up a little bit."
Speaker number one: "It did, because we pulled everything out of the market."
Speaker number two: "Thirteen cents a kilowatt-hour. That's excellent."
Speaker number one: "We pulled about 2,000 megawatts out of the market."
Speaker number two: "That's sweet. . . . Isn't that fun when you can do things like that now?"

— Traders for Reliant Energy ("Electricity from people you can trust"), June 2000

THE PAPIER MÂCHÉ TREE in The Barn looked like a Grade 4 art project. Its oblong paper leaves were attached with bright green pipe cleaners to a trunk water-coloured a dull brown. The whole thing seemed a bit hokey, and also a little out of place standing by a wall lined with boxes of serious-looking product specifications for motors, lighting, and refrigeration and power systems.

Each leaf on the tree featured a handwritten employee testimonial. Greg Pedrock's leaf said he simplified complex technical issues for customers. Elizabeth Chant's leaf indicated that she helped "to build stronger bridges to Vermont's affordable housing community—resulting in lower-cost, safer, healthier homes for low-income Vermonters, helping to design, develop and deliver the most innovative, low-income, multi-family program in the nation." At the top of the tree a proud, extra-large leaf proclaimed that Efficiency Vermont was the winner of the Harvard Kennedy School of Government award for Innovation in American Government in 2003.

Efficiency Vermont, housed in a converted warehouse beside Lake Champlain in downtown Burlington, is a utility with a difference. It pumps no water and generates no electricity. Although it acts in the public good, it is not a public utility. A subsidiary of the Vermont Energy Investment Corporation (VEIC), a non-profit corporation, Efficiency Vermont is an energy-efficiency utility, the first of its kind to operate statewide in the United States. In its first three years Efficiency Vermont reduced the need for energy production that causes pollution to the tune of over a million tons of carbon dioxide, the main greenhouse gas. It saved its customers $66 million in energy costs.

By the year 2000, just when Efficiency Vermont started advising farmers and ski-hill operators on how to save on their power bills, a practical notion that had been around since the 1970s was gaining widespread acceptance in both Canada and the United States: you can get more electricity, cheaper, by investing in efficiency on the other side of the meter than you can by sinking money into nuclear or coal megaprojects. Jobs could be created, and pollution reduced. The concept of "least-cost planning" had become *de rigueur* across North America by the late 1980s, when Ontario Hydro's Demand Supply Plan recognized that it made sense to build savings into the operations of the utility's customers, thereby reducing power demand, saving money, and cutting pollution. How to build those savings in was not so straightforward. Because least-cost planning flew in the face of traditional utility logic, it faced opposition from conventional utility planners and people working in engineering, construction, and fuel supply who had a vested interest in building and supplying power plants. As the story of Efficiency Vermont reveals, the keys are political will and regulation.

The "Green Mountain" state is famous for Ben and Jerry's ice cream and bucolic scenes of contented cows grazing beside covered bridges. Journalists inclined to label certain areas as "rustbelt" and "sunbelt" states might describe this one as a "granola-belt" state, and deploy images of greying back-to-the-land couples tending organic crops. Like all clichés, these descriptions carry a grain of truth that obscures more complex realities.

By 2004 Vermont had only one of every five dairy farms it had a generation ago. The state is now much more white and a bit more affluent than the rest of the country. It is also more sparsely populated, a largely rural state where well-off people from New York and Boston retreat to pricey ski-hill condos serviced by native Vermonters who often live in "manufactured homes" (formerly known as "trailers"). The largest industrial employer in the state is IBM; the company's major plant near Burlington is the Pentagon's primary source of computer-chip technology. The state has cute college towns with village greens, and it has forlorn villages and hardscrabble farms. Every two years Burlington sends its former mayor, the nation's only socialist Congressman, back to Washington. The state's stubborn populism is leavened with a good measure of local matters—and matters is a verb. A fellow in Barre started a place called the Farmer's Diner, trying to serve everything with ingredients produced within sixty miles of the kitchen.

Beth Sachs and Blair Hamilton, founders of Efficiency Vermont's parent company, are part of a loose network of alternative energy advocates that emerged from the radicalism of the 1960s and coalesced after the 1973 oil price shock. Hamilton studied architecture at McGill University, and he and Sachs both worked on solar- and wind-power experiments in Mistassini in Northern Quebec. They moved west to work for the National Center for Appropriate Technology in Montana and then went further west to California, where they helped start the first company to bring specially coated low-E (low emissivity) windows to a market eager for energy-efficiency improvements. They eventually settled in Vermont, where they started the Vermont Energy Investment Corporation. Initially a kitchen-table operation, VEIC became an agent for the state's housing authority. It started a housing loan program, averaging $4,000 for its energy-efficiency loans. It began doing energy audits, identifying wasteful practices and providing alternatives, just as big U.S. environmental groups such as the Natural Resources Defence Council and the Conservation Law Foundation were pushing the aggressive regulation that was least-cost planning.

Sachs and Hamilton, who had the technical expertise to provide testimony in support of least-cost planning arguments at regulatory hearings, had found a new market niche. They had the entrepreneurial zip to see that their services were needed when big consulting firms hired by "ious"—the unfortunate acronym for Investor Owned Utilities that, like Ontario Hydro, had got into big nuclear debts—were arguing the need for a new coal plant. ious like Vermont's Green Mountain Power resembled the old Ontario Hydro because they had a monopoly on power supply in a particular region. Unlike Hydro, ious in the United States tend to be privately owned. Another crucial difference is that they are subject to often stringent regulation by powerful public boards. Some twenty-two utilities operated, under regulation, across the state of Vermont.

"The universe changed when major environmental groups decided that the way to make things happen was to go to regulatory bodies and get them to recognize that power plants were huge polluters," explained Beth Sachs. "We were their technical pitbulls in court."

The arguments for least-cost planning were strikingly similar to those that had long been deployed by soft energy path advocates. (In fact, the people making both arguments were often one and the same.) What is the sense of arguing over electricity rates? Customers do not pay rates, they pay bills. Even if rates go up, bills can decline if less power is used more efficiently. In another reprise of the soft energy path, least-cost planners pointed out that people are not interested in electricity per se; they want lights to work and motors to turn. If you can find ways of getting lights to turn on and motors to turn over, and if you can simultaneously cut back on power bills and environmental damage, it would be absurd not to do so. It would be even more absurd to neglect such opportunities in favour of fossil fuels that contribute to global warming. This kind of thinking was gaining momentum across North America in the late 1980s.

According to Blair Hamilton, events in Vermont "pretty closely parallelled developments in Ontario," where the regulatory body was moving in the direction of pushing natural gas utilities onto the efficiency bandwagon. Although Ontario Hydro was still its own regulator, the environmental assessment into Hydro's Demand Supply Plan was hearing a lot about least-cost planning. In Vermont, with a stronger and more independent regulator, a landmark Public Service Board order in 1990 not only required all utilities to submit demand-side management plans but also laid down rules about how the utilities would estimate both the costs and the cost-effectiveness of the supply- and demand-side options. The rationale was relatively simple: utilities were already putting together resource plans for how they were going to meet future power needs and what mix of different power plants they needed to build; but they should also be evaluating the gains they could achieve through efficiency. They ought to treat gains through efficiency just as they would energy from a new power plant and buy as much of that as makes sense in the overall portfolio. In other words, utilities should be required to adopt conservation as an integral planning goal.

"If using less electricity on the customer's side of the meter is cheaper than generating it," Hamilton explained, "then that was deemed to be something that a utility should be required to do."

This approach implicitly recognized that the market's price signals would not by themselves produce a commitment to energy efficiency. After the 1990 regulatory order, Vermont's utilities began to search out the market for improvements in power-savings, hiring staff to promote savings to their customers. The results were encouraging but uneven; different utilities embraced the programs with differing levels of enthusiasm. Near Montpelier the Washington Electric Co-operative made the remarkable decision that it was not in the business of selling electricity. Originally established to serve a gravel-road countryside that was not sufficiently profitable for bigger commercial utilities, the Co-op began spending 10 per cent of its revenues on efficiency. Some eight years after changing its corporate goals, the Co-op members were paying more per kilowatt-hour but were buying fewer hours. Instead of going through a complex set of commercial transactions, the Co-op simply sent its contractors out with boxes of compact fluorescent light bulbs and gave them to householders after identifying the places where the less wasteful products would do the most good. The Co-op members were already paying for the bulbs on their bills. The Washington Co-op also embraced "fuel switching." As contradictory as it might seem for an electric utility to be urging its customers to start using natural gas or some other fuel source, the Vermont regulator pushed the electric utilities to do just that: the approach was seen to be in the best interests of the state—and of the state of the environment.

By the mid-1990s, when the Vermont Public Service Board's embrace of least-cost electricity planning was starting to pay off, another much more forceful wind was blowing across the U.S. political landscape. Planning was out and markets were in, at least according to Enron and the other companies that were pumping hot air into the deregulation balloon in the United States.

"We're doing more than signing a new law," a bold-sounding governor Pete Wilson said in 1996. "We are shifting the balance of power in California." This bit of prophesy on the occasion of the passing of the electricity deregulation bill displayed honest, if perhaps unintentional, foresight. Governor Wilson's message was that customers would now be in the driver's seat. Customers would now have the right to choose their electricity distributor. It was the biggest and most important experiment in electricity deregulation in the country. Previously the California electricity market had been a series of franchised monopolies regulated by the California Public Utilities Commission. Now power would shift from government regulators to electricity generators and marketing companies.

The idea of freeing business from the heavy hand of the state is hardly new. It is as old as capitalism itself; the roots go back to the era when a new mercantile bourgeoisie started to throw aside the rules of the rigid feudal system, replacing them over time with free-booting capitalism. It is a notion so ingrained in the discipline of classical economics that its fashionable revival under Margaret Thatcher and Ronald Reagan was known in the academy as neo-classical economic thinking ("neo-liberalism" to its opponents). Like nineteenth-century industrialists, its late-twentieth-century supporters indulged in ritual denunciations of the rigidities that resulted from state interference with a much more "flexible" free market.

California's neo-liberal electricity reformers took their cues from Thatcher's experiments in Great Britain, with its electricity market modelled on the flawed British Power Pool bidding system. But just as California got going, English electricity historian Walt Patterson was writing that power-station operators in his country had "routinely kept some of their plants shut down and unavailable, as part of their strategy to sustain a sufficiently high price in the electricity pool." But, he added, the operators had not yet seriously considered "the possibility of shutting down or not operating electricity facilities and letting the lights go out." Doing so would surely have resulted in public rage and government countermeasures.

At least at first glance, the move to deregulation in the United States seemed to spring from politicians like Pete Wilson responding to public pressure. Groups like Citizens for a Sound Economy (CSE), Americans for Affordable Energy, Coalition for a Secure Energy Future, Twenty-First Century Energy Project, and Consumers First! were just a few of the myriad organizations that emerged in the 1990s to tell anyone who would listen that the free market would provide a breath of fresh air, bringing lower prices and more flexibility to stodgy old power systems.

Speaking of public pressure: CSE commissioned a study proving that deregulation would reduce the average power bill by 43 per cent; it hired airplanes to haul "consumer choice" banners through the sky—which was all very well, except that CSE was a Republican front group and its research was funded in part by Enron. The democratic-sounding Americans for Affordable Energy included Ford and Enron. Consumers First! was a creature of the utility Southern California Edison, as was the Coalition for a Secure Energy Future, which claimed to represent "small and large business" (apparently in that order) as well as "taxpayers, consumers, educators, public safety, labour and ethnic organizations." The Twenty-First Century Energy Project was organized by Ed Gillespie, a George W. Bush campaign strategist who also worked for Enron, the company whose fortunes were inextricably bound up with the U.S. deregulation enterprise.

"I believe in god and I believe in free markets," said Enron CEO Kenneth Lay. Enron spearheaded a masterful campaign to loosen the rules governing the U.S. electricity market. It added the sale of electricity to its already established natural gas business in 1994, and by 1997 it was not only the largest electricity trader in the country but also the largest natural gas trader in North America and the United

Kingdom. The firm founded by Lay vaulted from being a medium-sized natural gas company in the early 1990s to become the seventh-largest company in the United States by 2000.

The Enron empire was built on faith in the good things that came from free markets for energy. According to *Newsweek*, Enron's executives were "deregulation's True Believers." The free-marketer's bible, *The Economist*, referred to Kenneth Lay (Kenny Boy to his friend George W. Bush, for whom he raised $700,000 in campaign donations) as an "energetic messiah. . . . Spend long enough around top Enron people and you feel you are in the midst of some sort of evangelical cult."

Not to be outdone, *Fortune* named Enron "America's Most Innovative Company" for five years running (1996–2001). In April 2000 the business magazine likened Lay's company to "the King" himself, bringing hipness to the squares:

> Imagine a country-club dinner dance, with a bunch of old fogies and their wives shuffling around halfheartedly to the not-so-stirring sounds of Guy Lombardo and his All-Tuxedo Orchestra. Suddenly young Elvis comes crashing through the skylight, complete with gold-lamé suit, shiny guitar, and gyrating hips. Half the waltzers faint; most of the others get angry or pouty. And a very few decide they like what they hear, tap their feet . . . start grabbing new partners, and suddenly are rocking to a very different tune. In the staid world of regulated utilities and energy companies, Enron Corporation is that gate-crashing Elvis.

Enron's $25-million nationwide ad campaign advocating retail deregulation was a key factor in the campaign to convince California to adopt retail electricity deregulation in 1996—the first state to do so. Enron was not alone in this. The Edison Electric Institute—an electricity industry lobby group with a long history—spent more than $11 million that year alone on its deregulation campaign; and other front groups were springing up across the country.

Unleash the magic of the marketplace, Enron preached, and competition will give you low-cost electricity. It was to be an exciting new era, even if no one seemed able to understand exactly what Enron was doing. What deregulated jurisdictions got instead was a high-rolling pyramid scheme that made the dot-com boom look like a flimsy scam by a low-level flim-flam man.

In the United States, given that most of the electrical utilities were already privately owned, the privatization enterprise was not so much about selling publicly owned utilities as it was about removing the forms of democratic oversight that governed their behaviour. During the 1930s, reacting to the excesses of the electricity "robber barons" of the 1920s and recognizing the public interest vested in electricity provision, President Franklin D. Roosevelt had established an extensive system of utility regulation. Part of the New Deal, the idea was to make sure that all Americans had access to the benefits of affordable electricity.

Even then Roosevelt had to overcome the pioneering efforts of the Edison Electrical Institute, which in its attempts to defeat the legislation used one of the earliest fake grassroots campaigns ("astroturf," as it later became known to practitioners of dirty political tricks). The Institute had been known as the National Electric Light Association until it was discredited by power industry holding company scams that prompted the original regulation of the private power industry. To oppose Roosevelt's legislation the Edison Institute sent politicians telegrams using names pulled from the phone book, at times delivering the messages at a rate of four thousand an hour.

Over time something close to a democratic process had evolved in which major decisions were made by public debate in an open forum with information available freely to all parties. In the land of free enterprise, government regulators capped private utility profits, told the utilities to reduce rates for the poor, and directed them to fund environmentally friendly projects to protect community employment: all of this in the context of open physical and financial inspection. By the 1990s the system still was not perfect, as the soft path types had been pointing out for twenty years, but for the customers prices were reasonable and service reliable. For the regulated firms, profits were low but predictable.

That was all so boring, and in Kenneth Lay's adventurous world, electricity was exciting. A kilowatt-hour of electricity could potentially be bought and sold ten times between the time it was produced and the time it was consumed. Of course, the physical electrons don't move between buyers and sellers—they move in response to the laws of physics—which is why it can be bewilderingly complex to try to track the resulting transactions when clever people are trying to cover their tracks.

⊕

Enron unleashed the full force of its lobbying and public relations machinery on California, and in 1996 its executives got the energy legislation they wanted with the introduction of Assembly Bill 1890. The new law allowed customers to choose the distributors they would buy from, but more importantly it forced the existing electrical utilities to sell off most of their fossil-fuel generating plants to unregulated private power companies. To make the sector more attractive to investors, the system now also included a charge on electricity to pay off the "stranded costs," whereby every customer helped pay down the debts associated with the cost overruns from building nuclear plants. To sweeten the pot for voters, the bill gave small customers a 10 per cent discount on rates and froze their rates until the end of 2001, or until the utilities had paid off their stranded costs—whichever came first. This cut was only a first taste, according to Enron and friends, who offered assurances that prices would tumble once competition was introduced. The market would work.

The legislation passed unanimously, and the California market was deregulated. For the first two years prices were stable, but by 2000 they had begun to rise. When the hot weather arrived the rates tripled, even though demand was generally lower. According to a study by the California Public Utilities Commission, the peak loads that summer were well below the levels in 1999, when price spikes did not occur. And even during the ostensible shortages of 2000, the state was exporting more power than in the previous year. Deregulation meant that there was now no way to meddle with this free-market activity. Enron left weakened government regulators feeling somewhat puzzled as the corporation played games with the market to boost prices and profits.

Maintenance suddenly became much more important for California's utilities; longer and more frequent shutdowns of generating capacity coincided with periods of high power demand. Sometimes a third of capacity was off-line. Before deregulation, utilities were required to explain the reasons for shutdowns. Under the new rules, or lack thereof, this was no longer the case. A company that owned generating facilities could simply turn one station off in order to boost prices by creating scarcity. Their other plants would make more money.

High prices were not the only problem. A new phrase had entered the North American political lexicon—"rolling blackout"—referring to a decision by the government or utility regulator to shut off all power in a certain area for a certain period of time. Normally, when it looks as though there won't be sufficient electricity production to meet overall demand, the government or utility regulator will cut off all power to one area of its jurisdiction on a planned basis, thereby reducing demand and stabilizing the system as a whole and avoiding a general, unexpected failure that might bring the whole system to a halt. When blackouts were ordered in California in 2001, hundreds of thousands of Californians groped in the dark. Oddly enough, the blackouts did not coincide with peak demand. One blackout even occurred on a Sunday in January 2001, when demand was only at 60 per cent of the summertime peak. The Ralph Nader consumer group Public Citizen pointed out that the blackouts hit when demand was at 30,000 megawatts, far below 1998's peak demand of 45,000 megawatts.

The whole free-market experiment ended when it was exposed as a massive swindle. A report to the state government by two deregulation supporters concluded: "There is considerable empirical evidence to support a presumption that the high prices experienced in the summer of 2000 were the product of deliberate actions on the part of generators or marketers controlling the dispatch of generating capacity to withhold supply and increase market prices." Later, in testimony before the U.S. Senate's Commerce Committee, David Freeman, the chairman of the California Power Authority, noted: "We must recognize that the so-called invisible hand of Adam Smith was Enron and their fellow gougers picking the pockets of Californians to the tune of billions of dollars. Prices were skyrocketing in California in late 2000 and early 2001 as a direct result of Enron's influence and participation." Freeman knew something about electricity, having run both the Tennessee Valley Authority and a local utility in California before taking charge of the CPA.

All along there had been plenty of generating capacity—the power was simply being turned off much of the time to force prices up. In August 2000 the utilities had shut down five times as much generating capacity for "maintenance" as they had done in August 1999. Plants that had been closed for maintenance an average of 5 to 10 per cent of the time for the previous five years were shut down 50 per cent of the time in 2000—which had led to those days when a third of the state's generating capacity was unavailable.

In September 2001 the ability of California consumers to exercise freedom of choice in power purchase came to an end; and just as there had been no surge of popular support for the birth of this much-advertised opportunity, there was little sign of public mourning at its passing. Fewer than one-half of 1 per cent of residential electricity customers had opted to exercise their new-found rights. The City of Los Angeles, which had not sold off its publicly owned system, had been immune to both price hikes and rolling blackouts.

The Big Lie swallowed by most of the popular press and political pundits blamed California's woes on inadequate supply, and some even blamed the problems on environmentalists who had blocked the building of new plants. The shocking thing is that anyone was surprised at what happened. The main purpose of electricity privatization is the introduction of profit-making into the system and the maximization of those profits. The manipulation of prices is hardly news to industry insiders. As R. Martin Chavez, a former head of risk management in energy trading at Goldman Sachs, told *The New York Times* after the Enron debacle hit the news: "The whole reason for the existence of traders is to make as much money as possible. I lived through this: if you didn't manipulate the market and manipulation was accessible to you, that's when you were yelled at." In a deregulated electricity market, supply margins are kept low so that prices and profits remain high. Generators can get five times the price per kilowatt-hour if they sell half as many kilowatt-hours. When the price spikes, consumers—most of them not even aware that the price has gone up until after the fact—just grimace and pay the extra fees.

Deregulation also meant huge staff cuts. Enron's CEO Jeffrey Skilling told an audience of electricity entrepreneurs to chop their labour forces: "You must cut costs ruthlessly by 50 percent or 60 percent. Depopulate. Get rid of people. They gum up the works." And cut they did. A 2001 study found that staffing levels in the electricity sector were down by 35 per cent compared to 1991. Those workers weren't just "gumming up the works"—they were undertaking necessary (but unexciting) maintenance work. Once they were gone, profits went up but inspections became less frequent and non-emergency repairs were deferred. Electricity workers pointed out the results—reduced system reliability and safety for both workers and the public—but to little notice, at least until the massive blackout in Northeastern North America in the summer of 2003. The Great Blackout was at least partly precipitated by a delay in trimming trees.

Enron, a company whose very name would eventually become synonymous with corporate crime, had risen to great heights by specializing in fast-buck commodity trading and hubris. It even described itself as "the world's greatest company." Through deregulation Enron showed that a company could "trade" electricity without owning a single generating station. All it needed was a market. "We believe in the economic benefits of open, competitive wholesale markets and we play a leading role in creating them," Enron declared. "Every day we strive to make markets." The way to make markets was to peddle influence, and no company was more actively greasing wheels and palms in Washington in the promotion of electricity deregulation than was Enron. In 1997 an ideologist from the Heritage Foundation predicted that deregulation would mean that "consumers, in free-market fashion, will call the shots in the electricity market, not regulators," and even blamed the power industry's poor environmental record on state regulation.

The focus in deregulated markets was on trading electricity as if it were a commodity, rather than on the nuts and bolts of keeping the system running. Susan Strange, a leading academic analyst of the deregulation of global financial markets in the 1980s, said that those interested in understanding the short-term behaviour of currency markets would do well to throw away their economic theories and study the psychology of young men high on cocaine. To understand electricity markets in the 1990s it would probably be worthwhile studying the psychology of young men in suits who have been watching too many *Star Wars* movies. Enron's clever traders called one of their favourite ways of cheating the system "Death Star" and even named one of the many shell companies they set up to disguise loans as profits after the *Star Wars* character Chewbacca. For the boys playing with electricity markets, it was all a game, and a very profitable one at that.

Death Star, Perpetual Loop, Black Widow, Red Congo, Cong Catcher, and Bigfoot: these were all code names for submitting fake reports indicating that large amounts of electricity would be shipped on a specific transmission line that couldn't handle the extra load. There was never any actual need for this electricity; Enron would accept a payment from the system operator to cancel the transaction once the system operator realized that there wasn't enough capacity to carry the load.

Another favourite trick was called round-tripping: two subsidiaries of the same company (or two companies colluding) would sell each other identical amounts of electricity to inflate the apparent volumes and hence revenue. If they both sell each other 1,000 megawatts at the same price, no money actually changes hands, but the result is that demand in the marketplace goes up by 2,000 megawatts, boosting the price for every other bit of electricity sold. "Get Shorty" and "Fat Boy" were other versions of filing false information to inflate projected demand and thus increasing prices.

To get around the price caps that had been instituted in California in a futile attempt to prevent price gouging, Enron and friends resorted to what they called "Ricochet," or "Megawatt Laundering." Given that imported power wasn't subject to the cap, a Californian generator would sell power to an affiliated company in the next state. The out-of-state company would then sell it back into the state at a higher price. The electrons themselves most likely never left the state. The most effective method of boosting profits, however, was simply to close down generators to create a false scarcity and thereby increase the price twenty- or fifty-fold above cost. In the standard explanation of the exercise of market power, it was all so simple, effective, profitable, elegant even.

All of this changed dramatically with the collapse of Enron and its subsequent bankruptcy. In November 2001 Enron shares fell below U.S.$1 from a high of over U.S.$90. The company that had pushed the hardest and profited the most from electricity deregulation would soon be synonymous with creative accounting, influence peddling, and outright thievery. By May 2002 U.S. regulators were reporting that they had found a "smoking gun" in the form of internal Enron memos outlining ways of manipulating California's power market. "About $30 billion was extorted from this state," said California governor Davis. "Those who claimed that there was no price manipulation here were just plain wrong."

The potential profits were huge. In the last two months of 2000 Enron made nearly $1 billion in trading profit in North America alone, including a staggering $440 million in trading profits from California in December 2000, the same month in which the state was declaring Stage 3 power alerts.

Much to the chagrin of market enthusiasts, only after the state government stepped in to close the spot market and launch aggressive conservation programs did the financial bleeding and blackouts stop. The results of the all-out push for energy conservation were particularly impressive. The programs cut peak demand by 8.4 per cent between the summers of 2000 and 2001. The first six months of demand reductions in 2001 saved Californians an estimated $660 million in spot-market electricity purchases and helped to avoid up to $20 billion in the projected costs of summertime rolling blackouts. Nor were the benefits purely financial: the combined effects of conservation efforts in 2001 and 2002 reduced pollution emissions by nearly 8 million tons of carbon dioxide and 2,700 tons of smog-forming nitrogen oxides relative to 2000. It was the equivalent of taking 1.5 million passenger vehicles off the road for an entire year.

◉

In the second half of the 1990s, the electricity privatization and deregulation express was just coming up to speed. By the time the era of "choice" in electricity dawned in California in 1998, twenty-two states with over 60 per cent of the U.S. population had either deregulated their electricity markets or were committed to doing so. But in Vermont a funny thing happened on the way to the free market.

The Green Mountain state was not immune to the deregulation trend that was sweeping the nation. A deregulation proposal that would have done away with the state's twenty-two regulated monopoly power suppliers, from big utilities like Green Mountain Power to the tiny Washington Electric Co-op, managed to make it through the state senate. As part of the legislative process leading up to deregulation, the Department of Public Service evaluated the progress that the monopoly utilities had made in energy efficiency. It found that having all those separate utilities deliver conservation programs had resulted in a duplication of effort and an uneven patchwork of programs across the state. "That was not unique to utility service programs," noted Blair Hamilton of the Vermont Energy Investment Corporation. "It's a characteristic of most markets." The Department recommended a new statewide conservation program, and the idea made its way into the proposed legislation.

The Vermont legislators also recognized that in any electricity system regulated essentially by the free play of market forces, the benefits of energy efficiency were likely to fall by the wayside. In the end the House of Representatives voted down retail competition. The state had looked at the supposed benefits of the free-market model being promoted by the national coalition of large power-users and Enron-style independent generators and marketers, and it essentially said, "No thanks." But the idea of the efficiency utility survived.

The state decided, though, that a single-purpose organization would deliver energy efficiency more efficiently. It put the job out to tender. The request for proposals called for a utility without turbines or poles and wires or meters, and Efficiency Vermont won the state contract. Although the state government would regulate the new undertaking and the traditional utilities would collect the system benefit charge that would fund it, the efficiency utility would remain separate from government and the franchised monopoly power suppliers—the idea of having the government do the job was politically unpopular. The arm's length relationship was underlined by having an independent trustee collect and disburse the funds.

Conscious that power-price increases were even more politically sensitive than the heavy hand of the state, the government tried to make the new charge as transparent as possible by stipulating that the efficiency charge would no longer be bundled into general rates, as it had been when the programs were being delivered by the utilities. It would be a separate line item on each electricity bill, a facet that presented the usual challenge to backers of such programs: when charges become visible, customers read them as a rate increase. As Hamilton put it, consumers would want to know, "If I'm paying this new charge, how do I benefit?" Energy-efficiency advocates faced a thorny communications problem. They would have to convince sceptical consumers that even if they saw an increase in their bills, the new system would end up lowering their costs in the long run.

Any success in delivering, and getting acceptance for, this all-for-one/one-for-all message requires at least a couple of elements: a political culture that includes a deep sense of public purpose; and a certain amount of political will. People have to accept the idea that doing more with less on the power front can benefit both the human species and every other creature that depends on a healthy environment. The shift in perspective does not necessarily require a dramatic, headline-grabbing crisis like the rolling blackouts that hit California after it fell for market liberalization.

Setting up an efficiency utility like the one in Vermont was "something anyone could do," said Chris Neme, VEIC's director of planning and evaluation. Before taking a big leap into the local conservation market by launching Efficiency Vermont, VEIC had been doing consulting work in Ontario for a decade. Neme had worked on regulatory measures designed to promote efficiency in Ontario's natural gas sector. By the time Efficiency Vermont won the Harvard award for innovative programming aimed at improving public life, it had been tapping what it called the "conservation market" for three years. It was offering schools and colleges $45 rebates on $180 VendingMiser devices that keep drinks cool but use motion sensors to turn off all non-essential functions when no one is standing in front of vending machines. With the rebate, it takes a year to pay off the investment in the VendingMiser. Installing the little control units to regulate electricity flow at forty vending machines at Middlebury College, where a program began in 2002, resulted in a decrease of 88,860 pounds of carbon dioxide emissions—the same amount of greenhouse gas produced by driving a car 96,000 miles.

Among the two thousand Vermont businesses that had worked with Efficiency Vermont by 2003 were dairy farms being pressured by agribusiness consolidation. When Sherry and Ron Machia of Sheldon decided to build a new barn for their five hundred cows, the utility showed up to help them out. "They heard we were building a large facility and told us about energy-efficient lights," Sherry Machia said. "A $14,000-a-year saving means a lot to us right now." The key to energy efficiency is to capture opportunities where businesses are making decisions, because the cost of conservation is far less when it is designed into a new structure.

At Fairbanks Scales in Vermont's Northeast Kingdom, a tri-county area just across the border from Quebec's Eastern Townships, the light fixtures were thirty-five years old. With the help of a $16,000 incentive from the state efficiency utility, the industrial-scale manufacturer installed a new $47,000 lighting system. The electricity savings totalled over $50,000 annually, a 100 per cent return on investment in less than a year for one of the largest employers in the area. "Business and the environment don't have to be at odds," said John Palermo, manufacturing and engineering manager at the company. "They can go hand in hand."

As part of the effort to sell Americans on electricity deregulation and mollify environmental critics, many states followed the Vermont example and included legislated supports for renewable energy and energy conservation as part of the restructuring package. Some thirteen states adopted Renewable Portfolio Standards (RPS) as part of their restructuring initiatives. An RPS establishes a minimum percentage of power on the system that has to come from renewable sources, and it increases this amount over time. California's legislation required that 18 per cent of the state's electricity would come from renewable sources by 2012 and 20 per cent by 2017. Governments justified such intrusion into the marketplace on the grounds that renewable energy sources such as wind or solar power are only more expensive than coal or nuclear energy so long as the sticker price of the dirty power doesn't have to pay for externalities—the costs they inflict on the biosphere and those who use it when they must pay for nuclear-reactor waste and decommissioning or have to breathe.

The regulatory requirement for a minimum amount of green power not only helps to level the economic playing field, but also ensures further investment in research and development for renewables. It helps to build the industry so that costs will come down in the longer term. U.S. renewable energy standards were, however, set at such a low level that they were little more than pale-green window-dressing on an overall package that, according to data from the Federal Energy Regulatory Commission, was resulting in increased coal use (because it was cheap, even if dirty) and the extension of the life of aging nuclear plants.

Deregulation had also undercut the U.S. utilities' conservation programs. The logic under the system that had been emerging since the 1970s was that regulated monopolies could (and should) compare the cost of building new plants with investing in conservation programs that reduced the demand for electricity. Under the new, deregulated system, utilities focused on selling as much power as possible, which meant that spending on conservation dried up. In the case of California, spending on conservation programs dropped from close to U.S.$500 million per year in the early 1990s to under $300 million per year in the 1995–97 period in anticipation of deregulation. Nationwide, the move towards electricity deregulation in the United States had cut spending on energy-efficiency programs in half. In response, the majority of states put a small surcharge—usually called a Systems Benefit Charge—on the price of every kilowatt-hour of electricity. These charges created a pool of funds that could be used to fund energy conservation, as well as low-income support programs, and research on efficiency and renewables.

Even in a political climate as apparently enlightened as Vermont's, conservation advocates have to keep good news stories front and centre because activist energy efficiency smacks of planning and tends to annoy interests that may be narrow-minded, short-sighted, or simply blinded by ideology. Despite Efficiency Vermont's early success and the eventual renewal of its initial three-year contract, it faced a concerted attack from Associated Industries of Vermont. In 2002 the manufacturers' association succeeded in curtailing a scheduled increase in the utility's funding, arguing that businesses should be left to themselves to make their own decisions about saving energy. Curtis Morin, president of an iron works, told the Public Service Board that he was dumbfounded by the plan for the increase in funding: "I cannot comprehend that anyone with the economic well-being of our state in mind would contemplate any action that would increase electricity costs." Yet if the full increase had gone ahead, the total charge would have amounted to two- or three-tenths of a cent per kilowatt-hour per customer, with businesses at the low end of the range.

Jane Whitmore, manager of appliance promotion for Efficiency Vermont, had the job of promoting Energy Star to consumers. Energy Star is a branding tool developed by government as a symbol of everything from light bulbs to houses that use less electricity. Whitmore, who had previously worked on Green Mountain Power's energy-saving programs, joined VEIC to help put together Efficiency Vermont's original efficiency utility proposal. The war-room-style map of Vermont on her office wall bristled with hundreds of coloured pins representing retailers who had signed up to offer Efficiency Vermont rebate coupons to people buying compact fluorescent light bulbs, room air conditioners, and lighting fixtures that met the Energy Star requirements. Having once sold five thousand compact fluorescent light bulbs at a single country fair, Whitmore was aware that the best way of getting people to keep on buying a better light bulb is to make your presence felt right where they usually go to buy bulbs. That meant deploying a small army of contractors to convince retailers to sign on with the programs. It meant training retail staff and maintaining point of purchase displays.

A former schoolteacher with a graduate degree in environmental studies, Whitmore brought an understated passion to the work. "When I was working with Green Mountain Power I was working to promote energy efficiency to their customers only," she said. "Now I am working for all Vermonters."

She waxed optimistic about the rise in public awareness of the Energy Star symbol. In the space of a year Efficiency Vermont saw a doubling of room air conditioners in store showrooms bearing the Energy Star logo. The California power crisis also helped. When she first started out at Green Mountain, compact fluorescent bulbs didn't come on right away and their buzzing and flickering tended to turn consumers off. But a crash program in California encouraged a huge market for the bulbs. Coupled with increased efficiency and the efforts of conservation programs, quality and price improved. In 1993 retail prices of compact fluorescents were $20 each and up. The products were hard to find because many retailers did not know about them. Those who did were often reluctant to devote shelf space to a high-priced item that might not sell. Some ten years later the average retail price was under $7, and the bulbs were carried at 130 retailers around the state.

For Whitmore, working for VEIC's founders—or "Beth 'n Blair," as they became known to their staff—proved rewarding. Every month eighty people would assemble for overflow meetings in The Barn to discuss their work. It was a comfortable place, with lots of mutual support and positive vibes about the "good stuff" they were doing. "Beth and Blair worked very hard from the start to make sure that the people who work here are people who really want to be here, who want to be doing this work," Whitmore said. "There isn't an appliance dealer in Vermont that doesn't participate in our programs. Or if there is, we don't know about them." Pausing, she added a caveat. Wal-Mart withdrew its four Vermont stores. "They didn't want to hand out coupons any more. It was just a bother for them."

Notwithstanding the apathy of the outfit with the tyrant's conceit of being the world's largest corporation, Efficiency Vermont enjoyed remarkable success. Vermont is home to some 600,000 people— about the same number as the Toronto suburb of Mississauga. By the time Efficiency Vermont had its first contract renewed, the utility had eighty-seven people employed to save electricity. In Ontario there was still no similar organization. Despite a supply crisis and the proven effectiveness of least-cost planning, conservation, demand-side management, energy efficiency—whatever the label—Ontario could not realistically say that it had as many people working to save power as tiny Vermont did. Around the time Jane Whitmore joined Efficiency Vermont, Ontario was, however, spending billions to set up an Independent Electricity Market Operator.

seven

All Aboard the Privatization Express: The Harris Tories

The status quo is not an option.
— political cliché

ON AN EARLY SUMMER evening in 1995 Mike Harris stood before a cheering crowd at a converted North Bay carpet outlet to tell his supporters where his newly elected government was going. Local Tory boosters knew their man as a frat-boy type not given to reading much beyond short briefing notes. So they were surprised when Ontario's new premier started spouting poetry.

> *Two roads diverged in a wood, and I—*
> *I took the one less travelled by,*
> *And that has made all the difference.*

The snippet of Robert Frost verse was supplied by one of Harris's handlers keen on showing that the new man intended to chart a bold new course for the province: cutting programs, getting government out of the way, slashing taxes; taking away the little supplement granted to pregnant women on welfare—they were supposed to use it to buy extra food but, according to Harris, they would just squander it on beer. In the election campaign, Tom Long, Deb Hutton, and the rest of Harris's team had successfully targeted the Liberals for being wimps when it came to welfare recipients and tax-cutting. Needless to say, Mike Harris's road less travelled by had nothing to do with a soft energy path, or anything else soft. It was all about making the choices that right-wing libertarians liked to describe as "hard."

Harris promised a Common Sense Revolution. Whereas politicians of all three parties had previously tried to use the government to fix problems, the new breed of common sense revolutionaries was adamant that government *was* the problem and that they wanted power in order to fix government. By fixing, they meant shrinking the size of the state and cutting red tape: making sure that the province was "open for business," and selling off everything that the private sector could do better than the government. The demonstrators were soon calling Harris "Mike the Knife," and the man who sharpened the blade for the premier was Maurice Strong's hired hand on electricity privatization, Bill Farlinger. In the years leading up to the Conservative election victory, Farlinger had helped Harris meet with some five hundred CEOs and corporate lawyers. This was the constituency that ponied up the cash that got Harris elected, first as leader and then as premier. Farlinger and five other businessmen personally guaranteed Harris's campaign debts, and immediately after the North Bay poetry recital Farlinger took charge of the Harris transition team.

Harris also turned to Farlinger when, like so many premiers before him, he needed a trusted ally to occupy the other seat of power in the province. The appointment of Farlinger as chair of the Hydro board of directors in November 1995 represented a changing of the guard. Adam Beck's Power for the People was being replaced by Adam Smith's invisible hand. The market would now govern Ontario's electrical system.

Farlinger, a career accountant specializing in corporate reorganization, admitted that he had no background in running a utility and that he was not familiar with the problems of its troubled nuclear system. As it was for Maurice Strong before him, shaking things up was his top priority. One of his first acts was to fire five Hydro board members affiliated with "interest groups"—unions, environmentalists, or opposition parties. Or, rather, he got environment and energy minister Brenda Elliott to fire them. The hapless Elliott, who had run an environmental store in Guelph, was as green as the Tory caucus got. She was left to explain how she wanted to reduce the size of the board as a cost-cutting measure. (Board members got an annual $5,000 honorarium and $185 per meeting, pocket change for a corporation with annual revenues of $9 billion.) One of the more polite common sense revolutionaries, Elliott took the trouble to thank the departing directors publicly for having "contributed a great deal to the province" through their work at the utility.

Power Workers' Union president John Murphy, one of the board members fired, commented, "The minister's claim that this move is for reasons of economy is the most feeble and pathetic justification for government censorship that I have ever heard." The real reason for the move, according to Murphy, was that the government was clearing out anyone who might oppose privatization. The Power Workers knew how poorly British electricity workers had fared when Margaret Thatcher restructured the power system, and they had no intention of taking this move lying down.

First, they went to court. The judge took Elliott's comments praising the fired directors at face value and thus determined that there was no just cause for their dismissal. He ordered Ontario Hydro to reinstate them. But the court case was only the opening volley in what became the most expensive public campaign by a union in Canadian history, financed by a surcharge on every PWU member. The union unleashed a $7-million ad campaign defending public power, using Niagara Falls as a symbol of Ontario's heritage. Bumper stickers sprouted up across the province saying, "Hydro Belongs to All of Us—Don't sell out, Mike!" Full-page ads in major newspapers screamed, "Watch out Ontario, You're About to Be Taken." The union pointed to Hydro's recent financial turnaround as well as the consequences of private power in the United States and Britain. It accused Hydro management and Bay Street of orchestrating "the biggest sellout in Ontario history" to line their own pockets. The campaign culminated in television spots that ran during the 1996 Stanley Cup playoffs.

The Power Workers weren't alone in their opposition. Clifford Maynes, an environmentalist member of the Peterborough Public Utilities Commission, said that the Tories wanted "to dismantle not just Hydro itself but the local utilities as well, and substitute the marketplace for planning in the public interest." He argued that the environmental community had long advocated the reform of Ontario Hydro. Many greens wanted some competition but most insisted on maintaining a publicly owned and controlled power pool scrutinized by a strong, independent regulator. According to Maynes, Farlinger's vision was based on "a minimalist model of regulation, in which the environment plays a very minor role." Maynes said the proposed nuclear privatization was "ludicrous" from a financial perspective— "unless what you're doing is privatizing the profits and socializing the risk."

But environmentalists were not united. Dave Martin of the Nuclear Awareness Project argued that environmentalists should remain "agnostic" about privatization, focusing instead on the regulatory structure governing any power company, private or public. He urged environmentalists to remember who their enemies were: Ontario Hydro, local public utilities, and the Power Workers' Union. "Our 'objective' allies in this power struggle are big business and the independent power people." The logic here was that because business wants cheap power, and gas-fired co-generation can provide it, the only thing that could prevent Ontario Hydro (or its successors) from re-tubing the reactors, or eventually even proposing a new round of nuclear construction, would be a solution based in the private sector.

For Energy Probe's Tom Adams, privatization would offer "a 'magic bullet' to solve our debt, rate, nuclear, and interest group politics problems." Energy Probe had been calling for the privatization of Ontario Hydro long before Farlinger or Harris had started considering it. The group's 1990 submission to the environmental assessment hearing for the Demand Supply Plan had been bluntly titled "Privatization of Ontario Hydro." Journalist Thomas Walkom noted that while "most Canadian public-interest groups support a greater role for government in solving environmental problems, Energy Probe calls for less. Its solution to the problems of the environment—and for many a related problem—focuses on free markets, privatization, selective deregulation and other policies usually described as right-wing."

Back in the early 1980s when Harris advisor Tom Long was a gadfly right-winger in the days of the Tory Big Blue Machine, Energy Probe saw itself as an iconoclastic outsider challenging the status quo of big government, interventionist social-support programs, and high taxes. But by the 1990s it was riding the crest of a free-market wave backed by the International Monetary Fund, the World Bank, and major multinational corporations. According to Energy Probe founder Lawrence Solomon, Margaret Thatcher was a bright and shining light: "No one in the history of the universe has better separated government and industry and so allowed for honest regulation." In one bit of biographical bravado, Energy Probe even described Solomon as a beacon in his own right:

> Following the publication of his findings in *Energy Shock, Breaking up Ontario Hydro's Monopoly* and *Power at What Cost?* the United Kingdom under Thatcher adopted his model in its 1989 reforms. Following the success of the U.K. privatization, jurisdictions around the world have adopted similar reforms, making Mr. Solomon's model the dominant model for electricity restructuring reform in the world.

Whether the power privatizers in London and Sacramento had sipped at the font of Solomon's wisdom is unclear. What is clear is that Energy Probe came to fulfil a unique niche: an environmental (latterly "consumer") group that the business press could quote approvingly as the privatization express gathered steam.

Notwithstanding the enthusiasm for electricity privatization from Enron to Energy Probe, Mike Harris's new government and Ontario Hydro's senior management still had a tough row to hoe. Even though they would succeed in dismantling (or "unbundling") the utility without much public opposition, at the start of their first mandate the Tories knew that they had to tread softly because the citizenry still regarded Hydro in a positive light. "There is a strong support for continued public majority ownership of Hydro," Hydro itself concluded. According to its own polling and focus groups: "Ontario Hydro is perceived as an important resource that people in Ontario feel they have 'grown and own.' They do not see it in the same light as other crown corporations, but as something that is literally part of their Ontario."

Such attitudes made things more than a little awkward for the managers of the Crown corporation. They were arguing that "there is no longer a strong rationale for complete public ownership of the electricity industry in Ontario" and predicting "significant benefits" from privatization. To make matters worse, the Power Workers' anti-privatization campaign had successfully moved the numbers, with public opposition to privatization rising from 52 per cent to 74 per cent. Tory strategists were left scratching their heads. One insider pondered the problem of how to describe the privatization of Ontario Hydro in the framework of a bumper sticker. "I understand welfare cuts," he said. "But I haven't figured out how you tell people what the benefits are of selling a public utility."

To try to answer this question, the government's Advisory Committee on Competition in Ontario's Electricity System released its report, *A Framework for Competition*, on the first anniversary of the Tory takeover. Harris had appointed the seven-member group at the same time as Farlinger was named chair of Ontario Hydro. The committee was headed by former federal energy minister Donald Macdonald. The Bay Street Liberal, an old friend of Farlinger, had also chaired Ottawa's big 1984 Royal Commission on the Economic Union and Development Prospects for Canada, which recommended a free-trade agreement with the United States and privatization of Crown corporations. He sat on the board of directors of at least three corporations (Siemens Electric, Banister Foundation, and TransCanada Pipelines) that stood to benefit from the opening of the electricity market. The corporate lawyer had mused that there was no "inevitability" about government ownership of Ontario Hydro. He also acknowledged that big users were the most likely to benefit from the opening of competition. Macdonald had described Canada-U.S. free trade as a "leap of faith." His advisory committee was now asking Ontario to make a leap of faith into a privatized power pool.

"We recognized Ontario Hydro's historic importance to Ontario, but we also believe that the province's existing electrical system is no longer compatible within the economic context in which Ontario operates today," Macdonald said. His "economic context" was a vague reference to economic globalization, which was painted as the backdrop for the inevitability of privatization and competition.

The committee recommended an end to Ontario Hydro's monopoly on electricity generation and transmission. It proposed that the province's sixty-nine hydroelectric plants and nine fossil-fuel plants be sold to private concerns. Because of "heritage concerns" (code for the success of the Power Workers' ad campaign), generating stations along the Niagara River would continue to be publicly owned, as would Ontario's twenty nuclear reactors. The committee members assumed that no one would want the Candus because of the ponderous debt load they carried. The reactors would, however, be treated as separate entities competing with each other in an open market. According to the plan, the province's distribution system would be consolidated into fewer utilities. When the marketplace was gradually opened to competition, first for wholesale and then retail customers, the Ontario Energy Board would be given new regulatory powers. The report offered assurances that a competitive electricity market would be all things to all people; it would lower wholesale electricity prices, elimi-nate the stranded debt, and generate tax revenues.

Yet, stung by the success of the Power Workers' campaign and the immense complexity of the proposed changes, the government let the report gather dust for a year and a half as consultations dragged on. Energy companies interested in buying Ontario electricity assets and investment dealers interested in brokering those sales carried out furious behind-the-scenes lobbying. The militantly anti-labour Tories also did some lobbying of their own, whispering sweet noth-ings to the Power Workers in hopes that the union could be brought on board.

Those efforts would ultimately prove successful. The union was already pro-competition, although still opposed to privatization. "The executive came to a decision early on that they'd rather be in the tent as part of the deal. They saw nothing wrong with competition, as long as ownership wasn't touched," said Allan Kupcis, CEO of Ontario Hydro at the time. The problem was—as Kupcis recognized—once commercialization was in place, what would be the "public policy reason" for hanging on to public ownership?

The publicly owned utility still had a good deal of solid if latent support. In the words that Tory education minister John Snobelen used on his own beat, the government needed to create a crisis if it wanted to make big changes. In the dog days of August 1997 it got its crisis, in the form of a scathing report on the state of Ontario's nuclear plants.

"Ontario Hydro's Foes Sniff Blood" trumpeted *The Globe and Mail*'s front-page headline on August 16. "Utility's sudden loss of credibility means it can't dictate policy the way it used to." The story below described how the provincial government was being pressed to open the province's electricity market to competition as a response to a damning report on the mismanagement of Ontario Hydro's nuclear program. Nuclear workers were soon being compared to the perennially addled Homer Simpson. Maurice Strong confessed that the nuclear division had been "in effect running amok" when he took charge in 1992, and he had never been able to get the information out of that section that he wanted. The usually elusive Farlinger made a rare public appearance at the news conference held to release the findings. He compared the corporation's nuclear wing to a "nuclear cult," and the media quickly fell all over the catchy sound bite. It was all a bit rich coming from the people who had so enthusiastically brought the nukes to Ontario.

The report in question had been commissioned by Kupcis, who had become Hydro's president during the NDP period, when Strong put him in charge of the unbundled utility's core business. Kupcis, subsequently CEO, was concerned about the persistent underperformance of Hydro's nuclear plants. At first he thought the problems were primarily managerial, but after an initial study in 1996 revealed much deeper problems Kupcis hired U.S. nuclear expert Carl Andognini to look more closely into matters. Kupsis explained the decision that would cost him his job in a way that hearkened back to the scientific nationalism of the original AECL-Candu project. "The genesis of the management accountability and the complacency so well documented in the Andognini report date back to the mid-1980s and grew. It was always 'we are the best in the business and always the best in the world and everyone came to look at how we did things and we never had to learn from anyone else,'" Kupcis said. "It became clear that we were in denial in our nuclear management of those problems and we were incapable of fixing them in terms of a root cause analysis."

A former U.S. marine with a reputation for turning around troubled nuclear stations, Andognini brought with him a team that came to be derisively known by Hydro staffers as the Magnificent Seven or the Dream Team. The Americans, who rode in wielding management texts instead of six-shooters, apparently exhibited little respect for the feelings of their northern cousins. After six months they presented Hydro's senior management with a fifteen-volume Independent Integrated Performance Assessment, which ranked the operation of Ontario's nuclear stations as "minimally acceptable."

The outside analysts found that Ontario's nuclear program— once considered amongst the best in the world—had a management culture characterized by arrogance, a tolerance for lax standards and inattention to detail, and a refusal to hear criticism. The culture did not encourage learning from mistakes or offer rewards for improvements. Rather, anyone who suggested changes was branded a troublemaker, or simply fired. The list of screw-ups was long and painful. Radioactive material was stored outside the fence at the Bruce plant. Workers were left in charge of control panels they weren't qualified to operate. Thousands of modifications had been made to the nuclear plants without safety or engineering approvals. Oil drums were stored directly below heating equipment. Radiation technicians had not received proper training.

The report, issued in August 1997, was not couched in conventional business-school bafflegab. It quoted Hydro employees saying disturbing things. "It is acceptable to cut corners." "The messenger with bad news will be told to fix the problem." "Not meeting commitments is the norm." What really made people sit up and take notice, however, was its conclusion: "Non-performance is accepted or even expected because senior management has neither set nor enforced standards. . . . As a result, few employees display all of the behaviors required to ensure that station operations remain within safety limits." This is not really what you want to hear about an operation that— if safety limits are exceeded—could well poison the population of the most densely populated part of the country.

Kupcis presented Andognini's report to Hydro's board of directors along with his resignation. Both of them were duly accepted. Kupcis— by all accounts one of the most pleasant, competent people ever to run Ontario Hydro—took what he later described as "kind of a sad elevator ride down" from the top floor of the Hydro building. Perhaps he thought that the goal of fixing the nuclear management problems would top the Hydro agenda and that others would pay a similar price for what had gone wrong. If so, he was wrong.

Within twenty-four hours the board had ordered seven nuclear reactors shut down so that the utility could concentrate on bringing the remaining twelve reactors up to snuff (one reactor at Bruce A had been closed in 1995, the casualty of a lead blanket left inside the reactor during maintenance). It also hired the U.S. consultants to fix the nukes. After Andognini and three colleagues got to work, resentful Hydro staffers burned U.S. flags and adorned reactors with stickers proclaiming "Candu, not Yankee do." By the time the Dream Team departed four years later the reactors were still not working, but the team members had collected $40 million. Although they had been in Canada only briefly, $17.5 million of that amount was for lump-sum pension payouts.

Meanwhile, the energies of the board and senior management were concentrated mainly on preparing the utility for the brave new world. According to Kupcis, the number one management priority after he left was opening up the market to competition and introducing privatization. The senior management and board left the "nuclear experts" who got the big bucks to mind the store, to take care of project management and rebuild the facilities. As David Brooks learned during his time on the Hydro board, the nuclear side had always operated without much scrutiny from the higher-ups. Now those higher-ups had other fish to fry. The government, which had used the Macdonald Committee to buy time, dusted off its report.

None of the information in Andognini's report was particularly surprising to Hydro's critics, who had long been pointing to the chronically expensive problems of the company's nuclear facilities. What was new was how the government proposed to fix the problem: privatize the plants and open up the electricity market to competition. "It appears that the chickens are coming home to roost," said Premier Harris in the wake of the Andognini report. For Harris, these particular chickens had been fattened up through inefficient government management. He added, "The whole state of the industry in North America is moving toward more competition. We cannot avoid that."

Three months after the release of Andognini's report, the government unveiled its plan for introducing competition in the form of a white paper, "Direction for Change: Charting a Course for Competitive Electricity and Jobs in Ontario." Echoing the Macdonald Committee report, the white paper cited four main factors driving change. The first was electricity deregulation in the United States, where federal regulators had mandated competition among wholesale electricity customers (municipal and rural distribution utilities) and where many states were passing legislation creating choice for retail electricity customers. The second was a concern for economic competitiveness. Even though Ontario had lower electricity prices than did most of its U.S. competitors, there was a belief in some quarters that deregulation would drive down the price of electricity there. A third factor was changing generation technology: smaller gas turbines were now capable of providing financially and environmentally attractive alternatives to large-scale, capital-intensive electricity generation. Finally came a concern about the debt associated with Ontario's public electricity system. This was the factor that a government trying to forge an efficient, cost-cutting image inevitably ran up the pole to explain its electricity policy. The implication was that private power would be lean, efficient, and more accountable.

In late 1997 Energy Minister Jim Wilson made the big announcement: "Today, we are flipping the switch on a more competitive and responsive electricity system and creating new opportunities for the people of Ontario. We're proposing to move away from a monopoly in generation and retail to a new era of competition."

That new era would, admittedly, be a confusing one. The minister was quick to add that the reforms and the industry itself were "complex" and "highly technical." But he was also reassuring. All would be well. "The bottom line is that these proposals would benefit every person in Ontario by restoring accountability and introducing the proven advantages of competition."

The new era of accountability did not extend to the nuclear restart program. The Tory appointees running Hydro were apparently so intent on preparing for the exciting new era of competition that they neglected the very real, but more mundane, problems that continued to plague the nuclear division. The original cost-estimate for bringing the four mothballed reactors at the Pickering station back on-line was $780 million. Yet the board and senior management of Hydro did nothing as the estimated cost for the job was raised ten times, and the completion date moved back twelve times. After four years (twice as long as predicted), the first of the four reactors finally began feeding power onto the grid in 2003, at a cost of $1.25 billion, almost three times the original estimate for that unit. The whole job—if the remaining three reactors could ever be made to work again—was estimated at a cost of somewhere between $3 billion and $4 billion.

⊕

It would take the Conservative government seven years and billions of dollars from the time it was first elected to introduce market competition to Ontario's electricity system. By the end of this period, as evidence from California, the United Kingdom, and elsewhere mounted, it was clear that the transformation was not only very complex, but also politically risky. Yet there was still no real proof of the advantages of competition. So why bother?

The Tories' free-market ideology provided a convenient rationale, but the real driver for electricity privatization came in its promise to make a few people very rich. This small group of potential beneficiaries was, of course, highly motivated, and its members proved adept at cloaking their private interests in the language of consumer choice, efficiency, and market discipline. These "interests" pushing for privatization were cut from the same cloth as the operators whom Adam Beck had in mind when he railed against "big interests." They fell into three categories: the users, the droolers, and the con men.

The users in this case were certainly not the residential customers, who were consuming about one-third of the electricity in the province. Average power-users simply want two things: lights that come on when they hit the switch; and a monthly bill based on affordable rates. Regrettably, the issue of whether or not the planet should be fried or poisoned to produce the power is not central to consumer concerns, particularly when the sources of that power are faraway central-station generators. Residential consumers thus have a marked lack of enthusiasm for the issue of "choice" in these matters. Quite sensibly, they tend to see little benefit in choosing a product that is, after all, only useful if it is exactly the same everywhere and always comes out of the same socket.

The big industrial users are another story. Echoing developments south of the border, a key force driving electricity privatization was the Association of Major Power Consumers of Ontario. The sixty-odd members of AMPCO were a who's who of Ontario's oil, chemical, automotive, mining, and pulp and paper industries. Regarded by Ontario Hydro as its "key stakeholders," they were using about one-sixth of the electricity produced in the province. For almost a century, Inco, Abitibi, Imperial Oil, and the other AMPCO members had expressed no ideological objections to the state-owned Hydro—as long as it provided them with cheap power. Their sudden conversion to privatization in the mid-1990s resulted from a heady mix of fear and attraction: they feared further nuclear-induced price hikes and they were attracted to the idea of deregulation as a way of reducing costs. According to Kupcis, when the big users saw the costs of Darlington being kicked into the rate base, they "switched from being our allies to being our enemies." The evidence coming out of the United States and other deregulated jurisdictions also seemed to indicate that big power consumers could use their market clout to snap up all the cheap electricity while leaving small business and residential consumers to pay for higher-priced power. If AMPCO members could use their muscle to grab long-term contracts with Ontario's water-powered generators, which were producing power at a cost of just over a penny a kilowatt-hour, the smaller purchasers would be left with the more expensive nuclear and fossil output—according to Hydro management's accounting, 4.3 cents and 7 cents per kilowatt-hour respectively. The big users would be able to reduce their costs dramatically.

Within a month of the Conservative Party's 1995 election victory, AMPCO had issued a position paper stating, "Ontario needs a competitive market for electricity if it is to remain on a level playing field with its international competitors." In 1996 the Association set up a front group to push for privatization. The Stakeholders' Alliance for Electricity Competition and Customer Choice included not only AMPCO but also the Ontario Chamber of Commerce, Canadian Federation of Independent Business, Electricity Distributors Association of Ontario,

Ontario Energy Association, and Independent Power Producers' Society of Ontario. The chairman of the Stakeholders' Alliance was former MPP and Tory party president David McFadden, who showed how privatization and personal gain can be intimate bedfellows: he served as counsel for the consortium that negotiated a lucrative deal to run and charge ever-higher tolls on the privatized Highway 407 north of Toronto. In its submission to the Macdonald Committee, AMPCO sounded a note that was at once threatening and cajoling:

> Failure to create a competitive electricity market will undermine Ontario industry's ability to compete in world markets and may result in the migration of capital and jobs to other more favorable jurisdictions. Alternatively, a bold move towards electricity competition will provide Ontario with a distinct advantage as the U.S. market struggles to deal with jurisdictional disputes.

It was indeed a bold move that the big industrial users were calling for: privatization of the transmission system as well as the generating stations, with everything to be sold off and the market opened by March 1998. They also provided a helpful critique of the positions of Ontario Hydro management (pro-competition, but wanting to move much more slowly and keep a large chunk of the old Hydro together to compete with U.S. utilities), the Municipal Electrical Association (pro-competition, but wanting to keep big users from cornering the market on low-priced power), and the Power Workers' Union (still opposing privatization). But their boldness had its bounds. Big industrial users such as Dofasco and Imperial Oil asked the government for, and received, a guarantee of below-market rates.

Then there were the droolers. "People are salivating all over the place," Hydro's director of Corporate Strategic Planning, Rod Taylor, said at the time. "And why not? It would be potentially the biggest issue in Canadian history." When the banks and investment houses looked at the electricity system, they did not see turbines and wires, cooling towers and containment vessels, but ten billion dollars' worth of infrastructure investments sitting around making no profits. For the financial industry, this was wasted money. Just imagining the potential commissions on the sale of the utility and the profits to be had from the privatized utilities had Bay Street all aflutter. The heftiest submissions to the Macdonald Committee were essentially "Pick Me!" proposals from would-be privatizers like the New York-based investment banker Morgan Stanley and investment broker Richardson Greenshields, an outfit swallowed up by the Royal Bank of Canada shortly after it submitted its report on how it could handle the financial end of privatizing Ontario Hydro.

Next came the con men. These were the new marketing companies—led by Enron—that had proven enormous profits were to be made from the manipulation of markets for essential utility services. The Harris government appointed Enron representatives to sit on its new Market Design Committee and to become members of the new Independent Electricity Market Operator.

Along with other private energy companies, Enron made the kinds of donations that got its executives' calls returned. After the 1995 election almost a million dollars flowed into the coffers of the governing Progressive Conservative Party from companies and individuals with interests in a privatized electricity market—and most of that amount came after the announcement of market-opening in 1997. The competition-friendly Liberal Party took in over $200,000 from electricity companies, while the pro-public power NDP managed $3,000.

The 1998 Energy Competition Act put in place the necessary statutory reforms to administer the new regime. The privatization enterprise had coined a new term, "unbundling." Some ninety years after Adam Beck had goaded it into creating the Hydro monopoly, the Ontario government broke Ontario Hydro up into five new successor entities. Ontario Power Generation (OPG) took ownership of all of the generation facilities, contrary to the recommendations of the Macdonald Committee, which had argued for breaking up the generating facilities into competing companies. To encourage the publicly owned power system to privatize itself, this new generation company was required by a proviso called the Market Mitigation Agreement to rebate to customers any revenue from prices in excess of 3.8 cents per kilowatt-hour. OPG would only be able to earn a higher price as it sold or leased its generating assets. The public power corporation was thus essentially intended to lose money until it offloaded at least 65 per cent of its assets.

Another unbundled company, Hydro One, took over the high-voltage transmission grid. Under the leadership of Eleanor Clitheroe, it immediately went on a buying spree, snapping up local distribution companies in preparation for taking on the big boys south of the border. The province's three-hundred-odd municipal electric utilities were quickly reduced to less than a hundred either through amalgamation or by being purchased by Hydro One. All local utilities were required to be incorporated as companies under the Ontario Business Corporations Act by the end of 2000.

The Ontario Electricity Financial Corporation was set up for the sole purpose of receiving income from the electricity business and using it to pay down the debt. The Corporation initially took charge of $20.8 billion dollars in debt, which was hived off from the nuclear plants and added to the ratepayer's account to make the nukes "competitive." This so-called stranded debt was to be paid off through a 0.7 cents per kilowatt-hour surcharge. Another body, the Independent Electricity Market Operator (IMO), was set up as a power pool to run the new electricity marketplace, and the Electrical Safety Authority took over some of the regulatory functions governing who could connect up to the grid and how.

Contrary to the phased approach adopted in Britain and elsewhere, competition was to be introduced simultaneously in the wholesale and retail sectors within two years. This showed a remarkable faith in the power of the marketplace and the ability of the sector to make the huge institutional changes required. As it turned out, the opening of the marketplace was repeatedly pushed back when it became clear that none of the parties involved were ready to make the move. Still, the introduction of competition was seen as a question of when, not if.

All of these changes, and proposed changes, were particularly vexing for the formerly non-profit municipal utilities that supplied power to most of Ontario. The utilities that had not been purchased by Hydro One were required under the new law to transform themselves into for-profit entities. The whole enterprise put huge pressure on the distributors, from the gigantic Toronto Hydro to the small outfits known as PUCS (Public Utilities Commissions) that dotted the province. Utilities that had been devoting their efforts to the mundane but vital work of maintaining power-supply systems found themselves adjusting to a completely new environment, with major institutional implications. Managers struggled with corporatization, marketization, new computerization, and changing regulation. The utilities suddenly had to start paying for lawyers to transform their corporate structures, but that was not the half of it. They faced tax liabilities for the first time. Their municipal owners expected dividends. There was much more red tape as they had to start conforming to all the standards of the new Independent Electricity Market Operator. The utilities faced large consulting costs related to putting into place the complex new computer systems that would allow distributors to buy power from the pool.

"You don't have to have a Ph.D. in economics to figure that the costs are going to go up," said Bob Lake of Peterborough's PUC in 2003. His company spent $2.5 million on new computer systems alone. It was an information technology consultant's dreamworld, a utility manager's hell. "Over the past five years nothing in the IT business has been on time and on budget," Lake said. (By 2003 the new Peterborough Utilities Group was generating a tidy annual profit of $4 million for its municipal owners. The city had wisely decided not to sell it off for $25–30 million.)

When the government passed its market-opening law in 1998 and told everyone to be ready within two years, the result was a combination of incredulity and anger. Lake did not understand the rush. "We thought they were nuts. Maybe they thought it was simpler than it was." Lake also told anyone who would listen that the market should be phased in gradually, first getting competition in generation working at the wholesale level and then ironing out the wrinkles before bringing in retail competition. But the Ontario government seemed determined to do it all at once, with a generation market still overwhelmingly dominated by OPG, a company it still owned. According to Lake, "You couldn't ask for a worse model."

In a huge paper and legal exercise, resources were diverted into dealing with the vagaries of market-opening. Although no one was quite sure what the future held, participants on the utilities side were painfully aware that if the lights went out or the price rose, there would be hell to pay. Consultants, though, had a field day as the looming free market in electricity promised a ready market for the services of companies specializing in electric power planning. "There must have been at least a billion dollars spent on the consulting sector over the past three years," George Davies, president of Acres Management Consulting and Ontario's former deputy minister of energy, said in 2003. "It was a rather expensive experiment in neo-classical economics."

A 1999 report from the Harris-appointed Market Design Committee outlined requirements for market participants, set out appropriate accounting and governance procedures for the Independent Electricity Market Operator, and suggested ways of preventing market manipulation. The Market Design Committee was dominated by people who stood to gain from the privatization enterprise, or at least thought they did. In addition to Aleck Dadson, the director of Government Affairs for Enron in Canada, it included representatives from the big industrial consumers, Brascan, the old Ontario Hydro, pro-privatization academics, and Energy Probe. Together they designed a system that had the spot market playing a central role, with all of the potential for abuses of market power that entails. Although the committee took its cues from the United Kingdom and California, none of the environmental protection measures that had been implemented in those jurisdictions found their way into Ontario.

Electricity is "deeply affected with the public interest." Those are the surprisingly passionate words of one economist who has studied how electricity markets work in practice, not just in theory. Electricity is the modern economy's oxygen—and is clearly to be distinguished from hog bellies or grain, commodities traded by speculators on the futures market. Generating and transmitting electricity are undertakings that cry out for planning.

The demand for electricity has little to do with price. Customers can't turn elsewhere for a substitute if the price jumps dramatically. They won't run home in the middle of the afternoon on a hot day to turn off their fridges when the price of electricity goes up. For one thing, they know their food would spoil. For their part, private energy producers have no interest in reducing power use, whether the energy is clean and green or dirty and dangerous. Across the United States, conservation spending plummeted as competition was introduced. During the run-up to the commercialization of electricity in Ontario, the Canadian Electrical Association, an industry group representing both public and private power companies, noted, "Investment in Demand Side Management (DSM) has been greatly reduced by electricity suppliers wherever a competitive model of electricity supply has been implemented." This lack of interest in conservation is not surprising. Traditionally, advocates of DSM have argued that cutting power demand means that integrated utilities would be able to delay or avoid entirely the construction of new generators. "In a competitive generation market," said the Electrical Association, "energy efficiency is no longer a reasonable activity for wholesale electricity retailers— who would want to sell less product?" What private power companies want is to sell a lot of electricity at the highest possible price, generating dividends for shareholders and bonuses for executives.

Between 1995 and 2001 power prices were stable as reorganization proceeded behind the scenes. The Harris Conservatives were re-elected in 1999, thus ensuring that the changes to the power system would continue, but the market-opening was delayed and then delayed again. It took environmentalists, a mixed crowd preoccupied with climate change and the fight for the ratification of the Kyoto Accord, over two years to respond to the Tory restructuring plans. Finally, in 2001, the greens came up with a *Green Report Card on Electricity Restructuring in Ontario*. "The government has failed on smog controls, failed on renewable energy targets and failed to promote energy conservation which is needed to keep consumer and public health costs down," said David Poch, a former Energy Probe lawyer who was commissioned to prepare the *Green Report Card*. "Ironically, the only environmental improvement from restructuring may be that runaway prices make people conserve energy."

Despite repeated assurances that the environment would be protected as electricity privatization proceeded, the Harris government's program lacked even the window-dressing that accompanied comparable moves in the United States, such as the introduction of Renewable Portfolio Standards. Oddly enough, it was the ostensible environmental representative on Ontario's Market Design Committee who led the opposition to Renewable Portfolio Standards in Ontario. Tom Adams of Energy Probe, appointed to the committee by Energy Minister Jim Wilson, argued that it was an unwarranted government intervention in the marketplace.

As a means of promoting renewable energy Ontario chose instead to rely on individual customers opting to pay a premium to purchase "green" power. In essence, the Tories were saying that if you wanted clean air or no nukes, then put your money where your mouth is and fork over an extra 50 per cent on your electricity bill. Similar green-power purchasing programs in other jurisdictions, even when well-promoted, had only managed to attract between 2 and 5 per cent of the market because, after all, those who pay more for clean power still have to breathe the same air as those that don't. Nor is there any evidence connecting the green electrons you've bought to the power delivered to your house—environmentally friendly electricity is just fed onto the grid in general. The connection between the clean power you pay for and what you get is so abstract as to tempt only the real keeners.

Trying to sell green power—where the individual pays more but everyone benefits—is an example of what academics call a classic "free-rider" problem. Economists and political scientists have traditionally agreed that a state needs government action to correct for market failure. In this case, however, the government was content to assume that people either have a lot of extra money or really like to feel like suckers. With residential users consuming only one-third of the power in the province, even an unprecedented level of commitment on the part of households would have a limited impact on the overall energy mix.

Although Ontario levied a 0.7 cents per kilowatt-hour charge to pay down the $20.9 billion in stranded debt, it added nothing to pay for conservation. The *Green Report Card* argued that because it was the nuclear plants that had created this debt, the debt-reduction charge was in effect a massive subsidy from the taxpayer to any private company that leased or purchased an Ontario Hydro—now OPG—Candu plant. Nor was this the only subsidy to private nuclear-plant operators. The government capped its legal liability at $75 million, an amount that would not even cover the legal fees in the event of an accident.

Private nuclear-plant operators were also shielded from the costs of decommissioning the plants once their useful life had ended. Ontario Power Generation had estimated these costs to be $18.7 billion for its three plants, but the environmentalists argued that this was probably yet another underestimate. In short, the private sector would get to keep the profits, while the public would be on the hook for the debt and the risks.

One of the side-effects of closing the seven reactors in 1997 had been a doubling of the output from Ontario's five aging coal-fired plants between 1996 and 2001, which meant twice the pollution. A quarter of the province's power was now being generated from dirty coal. The plants, long a major source of air pollution, were now producing between a quarter and a sixth of the various noxious compounds (such as sulphur dioxide and nitrogen oxides) that cook in the summer heat and produce the smog that blankets the southern parts of the province. Their share of the province's greenhouse gases production rose to 20 per cent. Ontario Power Generation's giant Nanticoke plant on Lake Erie became increasingly notorious as the single largest source of air pollution in the country.

Scientific evidence regarding the health impacts of air pollution was mounting just as the coal plants were gearing up. The Ontario Medical Association declared smog a public-health crisis in 1998 after its research found that smog was killing 1,900 people per year in Ontario—and that was using the conservative assumptions of the club representing the province's doctors. The body count did not faze the government, and the OMA commissioned another study. This one revealed that the hospital system was spending $600 million annually to treat the victims of air pollution, while the province's employers were losing $580 million per year because smog was making people too sick to go to work or forcing them stay home with their asthmatic children.

These statistics, together with their source, became gospel to Ontario environmentalists preaching the evil that was coal and the salvation that was energy efficiency and renewable energy. They started to do so with a born-again passion not seen since the anti-nuke movement of the 1970s and the campaign against acid rain of the early 1980s. By 2001 the federal government had been goaded into signing a smog treaty with the United States. The agreement called for cuts in pollution from Ontario coal plants far beyond what the Harris government was prepared to accept. This effort coincided with the emergence of the Ontario Clean Air Alliance (OCAA), headed up by economist Jack Gibbons, a former Energy Prober with a sense of hardball politics. He built the OCAA into a potent political force, with its membership of municipal governments, local utilities, and labour unions, as well as health and environmental groups, going beyond the traditional environmental sector. The Alliance would soon include such stalwarts as the octogenarian mayor of Toronto's sprawling western suburb of Mississauga: "Hurricane Hazel" McCallion boasted solid Tory roots and the giant 1,100-megawatt Lakeview coal genera-tor within her city limits. The single-issue OCAA remained neutral on the questions of power privatization and nuclear power (the Great Satan to many greens), concentrating on a single, simple objective: closing down Ontario's five coal plants.

Against a background of moribund nukes and filthy coal plants, Farlinger, Harris, and the rest of the privatization enterprise's true believers were running up against an apparently intractable political problem. Ontario had ample capacity to meet peak power demands, but with so much nuclear capacity down, perhaps for the count, the province was beginning to operate close to the margin. No longer was wintertime the peak period for power consumption. Instead, a vicious circle would begin each June. From the mid-1990s on, every summer seemed hotter than the last, making for more air-condition-ing use, more poisonous emissions from the coal plants upwind from Toronto, and more smog-alert days announced in the media (and in the political centre of the province). Yet the government was committed to selling off generators: otherwise the free market—when it finally did come to pass after the much-delayed market-opening—would not have its advertised effect of lowering prices. But who would buy power plants from OPG? And which power plants would they be interested in buying?

Private operators would certainly be interested in the coal plants. Even though their output had been doubled, they were still only run-ning at half their rated capacity, and the option was there to run them full out to take advantage of the higher prices that generators anticipated would come with an open market. Private operators could also expect to sell Ontario power into the U.S. market: international competition was another of the open market's much-advertised benefits. But given the summertime controversy of smog alerts, it would not look good for the government to sell the plants producing that smog to companies that would produce even more. The optics, as they say, would not be favourable, particularly with reporters chasing comment from environmental agitators who were already screaming blue murder on smog-alert days. Added to the mix was the mounting pressure to cut greenhouse gas emissions. Why sell a plant that should simply have been closed? The government, spooked, started to waffle on the issue of coal plants, putting a moratorium on their sale in May 2000 pending a review of the potential environmental impacts of privatization.

All of the polling and focus groups showed that Niagara Falls was so symbolically important to Ontarians that anyone who dared suggest selling the Adam Beck station risked the voters' wrath. Niagara was, as a result, also taken off the auction block, thus keeping the whole Hydro reform scheme off the political radar screen but complicating matters for the government. That left on the market the big dam at Long Sault on the St. Lawrence Seaway, along with a large number of smaller hydro dams or generating stations and the mammoth nuclear plants. Any pension fund or other institutional investor interested in reliable income—not to mention the growing independent power companies—would jump at buying OPG's gravy-train hydro plants. As long as it rains, the owners make money, big money. The cost of production—about a penny per kilowatt-hour—is extremely low. Even at the fixed price then in effect, the profits were both high and predictable. It was any investor's dream come true. But that, of course, was also an argument for keeping them in the public sector, which had long ago repaid the costs of building them in the first place.

OPG was desperate to sell something to comply with the terms of the Market Mitigation Agreement embedded in the Energy Competition Act, which limited the income that OPG could receive for its power until it sold off or at least leased its generating assets. The company had already broken up its generating assets into thirty-one legal subsidiaries, the better to privatize them. Eventually it did manage to sell one of the gravy trains. The 2002 deal transferring ownership to the Brascan Corporation of four dams on the Mississagi River, 100 kilometres northeast of Sault Ste. Marie, would be the only outright privatization of a generating station in Ontario. But the sale was not without controversy. Critics hollered about "firesale prices."

Brascan, an $11-billion conglomerate, had grown out of the Ontario-based Brazilian Light and Traction Company—one of the firms represented in the Electric Development Company, which had fought so hard against the creation of Ontario Hydro just after the turn of the last century. Now, at the turn of the twenty-first century, the company forked over $340 million for the four sites, with turbines capable of producing up to 488 megawatts of power, enough to power a city the size of Sault Ste. Marie. In the initial bidding, two U.S.-based companies had offered substantially more for the dams. Exelon Generation Corporation had offered $577 million, while NRG Energy Inc. had mentioned a figure of $381 million. Although these initial bids were not binding, they do suggest that Brascan's $340-million bid was low—and that it might not have been a good time to be selling electricity generating stations. Internal OPG records indicated that the Crown corporation would gain little or no economic benefit in selling the stations unless the price of power went down. If the price were to rise by 10 per cent, the stations would be worth about $380 million. (As it happened, in the eighteen months after the sale, the price would rise by 20 per cent.)

Although the rationale for the whole privatization scheme was that the government needed the money to pay down the old Ontario Hydro's much-advertised debt, the money received from Brascan did not find its way to the Ontario Electricity Financial Corporation, the Hydro successor company charged with debt repayment. Instead, The Tories dumped the proceeds into the province's general revenues, at a time when the publicly held debt was actually rising. This sort of offstage hypocrisy was not the only problem, as cottagers, outdoor enthusiasts, and tourist-related businesses in the Mississagi River area would discover in the summer of 2002. The dams on the river were producing power at a very low cost: about one cent per kilowatt-hour. Brascan quickly came up with a scheme of maximizing the profits: it closed the dams and let the water build up in the river and lakes behind them; then it opened the dams up and drained the system when the price was highest—often ten to twenty times the cost of production—which just happened to be in the summer. This approach wreaked havoc with the area's water levels and ecology, as well as with its important tourism industry. On Tunnel Lake, about 31 kilometres north of the Trans-Canada Highway, cottagers woke up one morning to see their boats beached. Further to the north, Rocky Island Lake turned into a muddy desert, with fish trapped in the remaining pools of water. Tourist lodges and outfitters were forced to close temporarily. The Tories were clearly not paying attention to watershed management.

Brascan was unrepentant. "We sold power when Ontario needed it most," said spokesperson Andy McPhee. "We were paid the same amount of money as any other generator was paid for electricity during that time. Next summer, we will again be operating our system within the legal limits, as we did this past year."

In the grand scheme of things, the Brascan deal was still small potatoes. The really big money was tied up in the nuclear plants, but the problem for would-be privatizers was imagining who would want one of the province's creaky, debt-ridden nuclear generating stations. No one in Canada had ordered a nuclear power plant in twenty-five years. The staggering capital costs of nuclear power make it dependent on government assistance and, along with billions of dollars in technical problems, these high costs had driven Ontario Hydro into the red. Conventional wisdom was that Three Mile Island and Chernobyl had thoroughly discredited the nuclear power industry. Just as Ontario was shopping its nukes around, *The Economist* lamented that nuclear power was so hopelessly dependent on public subsidies that any talk of a nuclear renaissance was misguided. The editors of that free-market bible even came out in support of a carbon tax to regulate greenhouse gas emissions.

Ontario, however, had been trying its best to make the nukes attractive to prospective buyers by stripping away the old debt associated with building the plants and handing it to the Ontario Electricity Financial Corporation. That move helped to make the plants more attractive, and several potential purchasers started kicking the tires. Eventually, OPG managed to hand the keys to the Bruce nuclear station over to British Energy plc, the big U.K. nuclear operator that was itself a product of Thatcher-era privatizations. It was the first major step towards privatizing Ontario's generating capacity.

⬤

"The opinion of financial houses internationally and domestically is that it was a good deal."

At least that was how Energy Minister Jim Wilson described the contract that the Harris government signed for the electricity-producing section of Bruce Power, the world's largest nuclear facility. Still, the government refused to release background studies by Goldman Sachs and the Canadian Imperial Bank of Commerce, saying that disclosure would be "injurious" to its financial interests; and the CIBC, whose chief economist was a key privatization insider, was not saying anything. Like all the other domestic and international financial houses, the bank was hoping for a fat payoff when Ontario eventually sold off any of its major power assets.

The government finalized the deal with Bruce Power in May 2001. The Bruce site had been home to the 1960s-era Douglas Point reactor, Ontario's first commercial Candu (closed down in 1984). That one was followed in the 1970s and 1980s by two four-reactor stations, Bruce A and B, as well as a storage facility for the low-level radioactive waste generated by the operation. Of course, neither the British engineers who took over the eight big reactors nor their bosses at home had any interest in nuclear waste. They would not have been at all interested in any deal that put a new operator on the hook for old waste. As for opposition charges that it was a bad deal, Wilson repeated the standard line, pledging that there would be no "fire sales" of OPG assets.

The Bruce deal was not a fire sale, for the simple reason that it was not a sale at all. Bruce Power, a subsidiary of British Energy, would lease the power plant from OPG and operate it until 2018, although the agreement had a bailout clause: the company could exit any time after 2006 if it decided that it was not making enough money. Needless to say, if prices rose after the market-opening, there was every chance that the company would stick around. OPG retained responsibility for the cost of nuclear waste management and disposal as well as plant-decommissioning (estimated at $7.5 billion for the Bruce reactors). Bruce Power's initial lease payment was $625 million. In addition, it was obliged to pay an annual rent based on revenue.

Included in the government's political calculus was Bruce Power's commitment to pay for the resurrection of Bruce A, a notoriously unreliable station that had been criticized in the Andognini report. The new arrival on the Ontario nuclear scene hoped to have two of the station's four reactors back on track by 2003. Once that happened, the province's power-supply crisis—electricity supplies were now tight, and Ontario was relying on power from the United States to keep the lights on—would be eased. The initial refit was pegged, perhaps rather optimistically, at $340 million, the same amount that the government would get from the sale of the hydro dams to Brascan.

Interestingly enough, the loudest voice against the Bruce deal was not an anti-nuclear activist or a Canadian nationalist but the former president of the American Finance Association, Myron Gordon. A University of Toronto finance professor, Gordon came out swinging against the deal right away. He knew all about transfer-pricing and the other ways that transnational corporations have of keeping subsidiary revenues down. Since the rent for the Bruce was based on revenue, Gordon was worried. "There are going to be 'anticipated future charges' and 'allocations of overhead charges' from England and all that jazz," he said. Then there were the numerous exit clauses. "If you read the contract, every other line describes how British Energy can walk away."

But Gordon's basic critique of privatized electricity boiled down to the consumers inevitably having to pay more for private power because their bills would reflect the return on capital demanded by private investors. He called the Bruce lease an "outrageous gift" to British Energy. "The Bruce has not been privatized," he said just after the contract was signed in 2001. "Only the profits have been privatized."

The profits would be higher if the price of power rose after market-opening, and like every other player in the new world of Ontario electricity, Bruce Power was preparing carefully for that momentous event. These players included the other generators (still dominated by OPG), the distributors (municipal utilities and the muscular Hydro

One, with its newly acquired distribution arms), other major power purchasers (Stelco, Inco, and all the others at AMPCO), and the new "resellers," such as Direct Energy, which were trying to pry individual customers away from their traditional utilities with long-term, fixed-price contracts. The new rules as set out by the Independent Electricity Market Operator ran to seven thick volumes. The new system operator was trying to cover all the possibilities of a system with a transaction flow that included millions of separate customers, companies running generators, retailers, and big users playing the spot market. It was the complexity of the new system that caused so many delays and provided so much work for pricey consultants.

"Inevitably there is political sensitivity about opening the market because of what people have seen happening in California," said Andrew Johnson, who had arrived from England with other British Energy operatives to run Bruce Power. Johnson was vice-president of power marketing for the new company on the Huron shore. Along with everyone else in the industry, he watched in fascination as America's highest-profile effort at deregulation turned to ashes.

What happened in California "must not and cannot happen here," he said in the summer of 2001.

> What you do by going through deregulation is create very strong forces driving [electricity] prices downward. You can't say in one month this is going to happen. You can't even say that in one year. But if you look over a period of time, what you can be sure of is that prices are going to be lower than they otherwise would have been.

Which was, of course, impossible to prove one way or another. But to get an idea of where things would be heading, the government simply had to give the final okay to market-opening. It was up to Mike Harris to get the province finally rolling down the road less travelled—the one he had ruminated on so poetically some six years previously.

eight

Danger Ahead: The Wheels Fall Off

Today we have a situation in Ontario where there is no planning, and that's what we're suffering from.
> — Lorne McConnell, former head of planning
> for Ontario Hydro, 2003

ONTARIO'S PREMIER SAW himself as a no-nonsense kind of guy, a straight shooter. He was fond of talking about "a promise made" being "a promise kept." In mid-December 2001, having announced his retirement several weeks before, he was perhaps looking for proof positive that he had kept his 1995 promise to privatize something.

He had not managed to sell off TVOntario or the government liquor stores; and it had been over six years since he had picked his political mentor, Bill Farlinger, to oversee the privatization of Ontario Hydro. Since then Ontario Hydro had been divided up, and there had been much talk about free-market power. In late 2001 Harris finally did something decisive about Ontario's electricity system. He made two important decisions. One was expected, but the other was more of a surprise.

The surprise announcement was that Hydro One, the huge transmission company that had been created with the unbundling of the old Ontario Hydro, would be sold to the private sector. It would be the biggest privatization in Canadian history, and the proceeds would go towards paying off the debt left over from the manic nuke-building spree. As it turned out, over $100 million in commissions and fees would line the pockets of Bay Street brokers whose companies would handle the Initial Public Offering (IPO). It is unknown whether Harris expected any lucrative directorships as a result of this happy payday for his party's traditional backers. What Harris did make clear was that the future for electricity in Ontario would be rosy. He even made one of his famous promises, pledging that the government "will continue to regulate the energy sector to ensure that the market is fair, make sure it is open, and make sure consumers are given the best price possible."

A few days later Harris was in front of the cameras again. After four years of meetings and millions of public dollars being spent, he had an announcement to make. The market would finally work its magic for Ontario's electricity system, starting on May 1, 2002. The Independent Electricity Market Operator (IMO), staffed mainly by former Hydro employees, was ready, and so too were most of the local distributors who sold power to households and businesses across the province. The only joker in the deck was the state of Ontario's aging nuclear power plants, still undergoing an endless refitting job. After years of massive expense, it was still uncertain as to whether they would be able to produce enough power to keep the lights on and the price down. That's why "market-opening," as it was known among those who had been waiting for it for so long, was slated for the springtime shoulder season, the time in between peak periods when power demand is moderate.

The outgoing premier sounded a reassuring note. "Nothing," he promised, "is going to go wrong."

Later, on the first anniversary of market-opening (and with the benefit of hindsight), CBC-Radio's John McGrath could not resist playing those six words three times. But right after Harris made his announcement, another journalist, the former anti-Darlington campaigner Paul McKay, had predicted in *The Ottawa Citizen*, "Harris has lit the fuse on a fiscal and political timebomb, [and] then done his damnedest to make sure it detonates on someone else's watch." Indeed, by the time things went horribly wrong, Premier Harris was gone and his successor was left to deal with one of the biggest political trainwrecks Ontario had ever witnessed.

The backers of the experiment in free-market electricity had long been urging the Tories to get on with it and not to worry about what was happening south of the border. Earlier in 2001 economist John Grant had stressed that although prices had been "volatile" after California's market-opening, Ontario had little to worry about because the problems down there had been caused by a lack of new generation—the lie frequently repeated and swallowed by many media outlets at the time. Grant was a restructuring insider who represented the private interests and ideology backing Ontario's restructuring project. A former member of both the Macdonald Advisory Committee on Competition in Ontario's Electricity System and the Market Design Committee, Grant was on the board of the IMO. As chief economist of CIBC World Markets, he was fond of spouting shopworn metaphors that telegraphed the ever-reliable nostrums of neo-classical economics. "The bottom line is, we'll probably avoid the excessive price volatility experienced elsewhere. But consumers will have to be more responsible, since empowerment brings choices."

Within weeks of Harris's announcement of market-opening, the government was running ads telling people they could stick with their electricity supplier or sign up with a reseller. "The choice," the ads concluded, "is up to you."

Aside from the big industrial customers at AMPCO who had agitated for the breakup of Hydro from day one, most of Ontario's electricity users were unaware that they were about to be empowered by the market. It had not been their idea. There had been no mass campaign to liberate electricity buyers. Although a new breed of resellers had convinced some people to sign on for fixed-price contracts, most stayed with their traditional municipal distribution companies or with Hydro One, which was emerging as a major distributor. Hydro One management under CEO Eleanor Clitheroe and chair Sir Graham Day had been lobbying hard to have their company privatized.

Plans for selling off Hydro One had been afoot, in a quiet way, for some time. It was understood, but usually unstated, that executives like Clitheroe and Day stood to make more money—much more money—when the privatization went through. What was not so well known was that, under Clitheroe and Day, Hydro One had hired the services of a cabal of Tory insiders to sell the privatization idea to their friends in the Harris government. The backroom operators received untendered contracts for a secret, eighteen-month project that bore fruit with Harris's surprise announcement. Paul Rhodes, a former spokesman for Harris (and someone with no utility experience), received $335,237 over eighteen months for advising Day on how to butter up the premier. Rhodes's ignorance was such that he claimed in a memo in early 2000 that the transmission utility was in a "death spiral." At the time Hydro One was the envy of investors lining up to buy it because of its steady returns. What Rhodes didn't know about finance he made up for in his apparent insights into Harris's character. Several months later Rhodes advised Hydro One that Day should "stroke" Harris by telling him he was the "first premier to grab Hydro problems and do something about it." Coming from his long-time spin doctor, the suggestion that Harris would be so susceptible to flattery was either highly insulting or very revealing.

Hydro One also retained Leslie Noble, another Tory insider. She did some editing on Rhodes's risible "death spiral" memo, wrote a three-page memo on the pros and cons of privatizing the company, and put together a couple of slide presentations. For this she received a quarter of a million dollars. Tom Long, who was at the time campaigning against Stockwell Day to lead the Reform Party, co-authored a memo suggesting that Hydro One's shares should be deliberately underpriced so that private investors could make some fast money when the new stock rose after the sale. Long worked on the memo with Harris's former director of policy, John Toogood. Together they concluded that fattening wallets on Bay Street was more important than "delivering an extra few hundred million to the Treasury." Two of Long's little firms got $1.3 million from Hydro One during this period.

Insider sleaze aside, there was an elusive aspect to the whole electricity makeover, arguably one of the most far-reaching and important policy shifts Ontario had witnessed in decades. The biggest Crown corporation in Canada, a utility that had symbolized a sense of public purpose since the days of Adam Beck, had already been taken apart. Part of its generating capacity had already been leased to a private British company. Much of the rest was slated for privatization under the market mitigation policy intended to reduce the power of Ontario Power Generation. Now a brand-new market in electricity was about to be opened up. Hydro One, the publicly owned business wired into every generator and most every customer in the province, was to be sold off. The privatization of Hydro One would do nothing to foster competition because the company had a natural monopoly and no one was ever going to build a rival grid. In fact, Hydro One was a low-cost transmission company that promised to do nothing but deliver reliable power to customers and a steady, reliable flow of profits to its owners, whoever they were.

What made the whole thing so elusive was that all of this had been taking place without much public comment, as power bills had not risen and the government had shied away from privatizing any high-profile facilities. To be sure, the banks, consultants, electricity resellers, and potential generators who stood to gain most directly from the enterprise had been vocal and supportive from the start, with market enthusiasts like the CIBC's Grant offering advice. Legislators, however, had not been deluged with letters from their constituents demanding that they be allowed to choose their power supplier, or suggesting that Hydro One be sold.

However, it is equally true that there had been no upsurge of support for Hydro. The historically minded knew about how the shrewd Adam Beck had nurtured a populist movement that had mobilized Ontario against the Big Interests represented by the Toronto financiers who would surely have gouged The People if they were permitted to run Niagara's power and control the transmission towers that marched away from the falls. By 2001 Beck would have been turning in his grave if he had been able to see the impending sell-off of the people's generators to the likes of British Energy. If he had known that even the transmission system was on the block, he might have been spinning fast enough to generate a bit of electricity himself. Despite it all, the political silence that greeted the long years of preparation for Hydro commercialization had been so deafening that veteran politico Floyd Laughren was taken aback. By this time Laughren was chairing the Ontario Energy Board, which gave him an inside seat at the Hydro restructuring table. But even earlier on, under the Harris Tories, as a member of the legislative committee that toured the province investigating the woeful state of Hydro's nuclear plants, Laughren had easily spotted the public resentment, indeed animosity, against the public utility. The public enterprise sentiment that once might have supported the former Ontario Hydro had dissipated.

It had all been very easy for the Tories to do what they were doing, he explained before Harris announced the Hydro One sale and the date of market-opening. "In all my years of public life I have never witnessed so little comment or resistance to such a massive public policy change."

On the political right, most of the business class was behind privatizing both the wires and the generators and injecting market relations into the wholesale and retail power sectors. In the days of Beck an Ontario manufacturing bourgeoisie had backed public power because they wanted an ensured supply of cheap electricity to run their factories. Now the major power consumers at the influential AMPCO were abandoning their long-time support of Hydro. Whether this was because they really did believe that free-market electricity would be cheaper, or whether it was a case of ideology trumping self-interest, was unclear. Perhaps it was because ideology and self-interest could march hand-in-hand if guided by political influence: in 2000 the Harris Tories had ordered Ontario Power Generation to extend the sweetheart deals that fifteen or twenty big corporate power-users already enjoyed. These outfits would continue to get discounted power for four years after market-opening—about 8 per cent of the province's peak electricity demand was being sold to companies at cut-rate prices. But the government kept the names of these lucky companies and their preferential rates secret; it provided Ontario Hydro's successor companies with a special exemption from Freedom of Information laws.

A few cracks began to appear in the business common front when Hamilton steelmaker Dofasco began to balk at Hydro One's ambitious plans. As soon as the company was created, Clitheroe and her management team had gone on their ambitious buying spree—spending over half a billion dollars to buy up eighty-eight municipal utilities. Hydro One was girding itself to become a major player in a newly deregulated U.S. power market. Dofasco public affairs director Gord Forstner, who also had experience on the board of a municipal utility near Hamilton, worried about having the steelmaker's competitive edge threatened by New York heatwaves. "If we've enjoyed a competitive advantage because we've got lower-cost electricity, why would we want to give away our competitive advantage?" he asked.

Soon after the Hydro One sale was announced, Forstner's boss described the Tory privatization strategy as "half-assed."

Veteran Hydro-watcher Sean Conway, who had taken over as dean of the legislature after Laughren left, put it bluntly: "The business lobby and AMPCO were the Revolutionary Guard of this policy. But some of the first bits of medicine they got, they didn't like." Still, the government's commercialization moves had ample business backing from the financial and energy sectors. The juggernaut seemed unstoppable.

Over on the left, there were few guardians, revolutionary or not, defending public power—that is, until a man wearing a top hat and a swallowtail coat threw a big, old-fashioned switch at the corner of Queen Street and University Avenue. The Adam Beck photo opportunity was staged in front of the memorial to the Hydro founder just before Premier Harris revealed his privatization and market-opening decisions. It was intended to launch a new organization called the Ontario Electricity Coalition. Standing in front of a giant lightbulb emblazoned with the slogan "Save Public Power at Cost," Paul Kahnert laid out the left's argument: there is no way private power can be cheaper than power from a public utility, because private firms need a profit, which would inevitably get added onto utility bills. Early in a campaign that the Coalition promised would reach into every corner of the province, Kahnert, a Toronto Hydro line-crew foreman, showed he could make a sharp point. Privatizing the generators and wires that had cost the public so much money was silly, he said. He pointed to a line item called "Debt Retirement Charge" that was now appearing on every power bill. "Does it make sense to be paying the debt on assets we no longer own?" he asked.

Such critics had a major problem: they had to position themselves as supporters of public power without seeming to defend the old Ontario Hydro. The former state monopoly's public-power-at-cost imperative had long been criticized—particularly by environmentalists—as far too costly in ways that went beyond Hydro's balance sheet or customers' bills. The broader costs included acidified lakes, deadly nuclear waste, and global climate chaos. The trade union movement had never been onside with the green movement's anti-nuke campaign, in good measure because the Power Workers' Union held a virtual veto over any official labour backing of traditional anti-Hydro (read, anti-nuke) campaigns. But by this time the PWU—Local 1000 of the Canadian Union of Public Employees (CUPE)—had blotted its copybook with labour by supporting the hated Harris government's privatization policies in the electricity sector.

In doing so, the union local was increasingly out of step with important parts of the labour movement. Environmental and labour activists had recently gained a practical experience working together, not only during the fight against the government's attack on social programs, but also on cancer prevention campaigns, the issue of banning cosmetic pesticides, and "just transition" programs to compensate and retrain individual workers if environmental rules brought job loss. In a politically important departure from developments south of the border, Canada's labour movement stood with environmentalists in calling for the ratification of the Kyoto Protocol on climate change, with the Communications, Energy and Paperworkers union leading the charge. One reason the energy industry's massive campaign against Kyoto in 2001–02 flopped was because of its failure to frighten its own employees. The workers could not be persuaded that their bosses were being anything but selfish and short-sighted.

CUPE as a whole was, not surprisingly, especially adamant about protecting public services and public-service jobs and had started to put money into the Ontario Electricity Coalition and its campaign. But on the question of privatizing Hydro, the greens seemed just as divided as trade unionists. The Coalition included the Toronto Environmental Alliance, but prominent anti-nuke campaigner David Martin, by this time with the Sierra Club, immediately reacted to the OEC's Adam Beck launch by recalling the bad old days of the Ontario Hydro monopoly. He warned about the "big mistake" of getting "trapped into an ideological debate about ownership."

Yet matters ideological had always been inseparable from matters electrical. In the early twenty-first century, who could have been more ideological than Jeffrey Skilling? Skilling had just resigned as CEO of Enron, the world's leading corporate apostle of private ownership of electricity. He was, according to *Newsweek*, "like a religious zealot who couldn't stop repeating his favorite mantra as the solution to all the world's problems": deregulate. When Kahnert helped launch the campaign against Ontario's privatization enterprise, he was holding a hand-lettered sign with the headline "Deregulation?" Under it were five capital letters: Electricity Nightmare Ripoff, Ontario Next?

🐘

The unveiling of the extent of Enron's manipulation of electricity markets and its contribution to California's crisis in 2000 couldn't have come at a worse time for the Ontario government. Around the same time that Harris was announcing the opening of Ontario's market and the selling off of Hydro One, Enron—which had helped to design Ontario's market and had been the biggest booster of electricity deregulation in the world—collapsed in the largest bankruptcy and corporate scandal in U.S. history. Just as the Ontario government was trying to convince an increasingly sceptical public that entrusting the electrical system to the private power companies was a good idea, U.S. regulators were lining up to get back some of their citizens' money.

The shady dealings surrounding electricity policy changes in the United States did not go unnoticed in Ontario. By the winter of 2002 the ruling Conservatives were in the midst of choosing a new leader and the public was finally becoming aware that big changes were in the works with the impending market-opening and privatization of Hydro One. The company had netted $978 million in profits in its first three years of operation. It was a busy time for the proponents of privatization.

The Independent Electricity Market Operator, together with the underfunded Ontario Energy Board and Ottawa's toothless Competition Bureau, agreed on each other's oversight responsibilities in ensuring that companies participating in the province's open electricity market would not abuse their market power or engage in anticompetitive pricing or other monopolistic practices. The Bay Street bankers and lawyers who had lobbied so strenuously for privatization busied themselves with the preparation of a massive Hydro One Initial Public Offering that looked as though it would be twice the size of Petro-Canada or Canadian National Railway. The transmission monopoly's regular earnings meant that the underwriters knew the privatization would be an extra-easy sell, particularly to cautious investors looking for predictable dividends. Indeed, it was so juicy that the prospective underwriters even agreed to lower their commissions. On a $5-billion deal the Royal Bank, Bank of Commerce, Toronto Dominion, Bank of Montreal, and Bank of Nova Scotia could still look forward to sharing in a $113-million payday. The money would be shared with Goldman Sachs and a raft of much smaller securities dealers, but the big banks would fare the best. With the consolidation of the financial sector, they now controlled the companies that dominated the underwriting business.

Bankers are always anxious to cultivate political contacts at the highest levels. Former Finance Minister Ernie Eves, the leading candidate to replace Mike Harris, had gone directly from cabinet to the executive suite at Credit Suisse First Boston. That company was part of the "global sales team" organized by investment dealers to handle the Hydro One IPO. Brascan, the company that would buy the money-spinning hydro dams in Northern Ontario weeks after the leadership contest concluded, donated $150,000 dollars (more than three times the next highest corporate donation) to Eves's successful drive for the premiership.

The Stakeholders' Alliance for Electricity Competition and Customer Choice continued to back market-opening and privatization, but the cracks in AMPCO's common front widened as reality began to intrude on ideology. Dofasco, Ontario's largest industrial electricity customer, was still worried about the dominant position of Ontario Power Generation. The steelmaker realized that, given Hydro One's desire to enter the U.S. market (in late 2001 the utility had applied for approval of its plan to lay huge transmission cables under Lake Erie to increase integration with the U.S. power grid), the future looked cloudy, particularly for its electric arc furnaces. The way things were shaping up, predicted Dofasco CEO John Mayberry, customers were going to be slapped hard. "Prices will go up and consumers will have little ability to protect themselves," he wrote.

> We will pay for generation. We will pay for transmission. We will pay for distribution. In the new system, at each step of the way, a profit needs to be made by providers. Then we will have to pay for the stranded debt left over from the former provincial power sup-

plier, Ontario Hydro. We will pay for a market operator to make all the rules and handle all the transactions.

In a last-ditch effort to hold off the Hydro One privatization, some of the big power-users tried to convince the government to turn the wires business into some sort of non-profit entity. In this they had the backing of Tory bagman Tony Fell, who also happened to chair RBC Capital Markets, the biggest broker on Bay Street. They pointed out that a privatized Hydro One would have to start paying taxes and that some of those taxes would go to Ottawa instead of staying at Queen's Park. In addition to making this "leakage" argument, the cautious elements of the business class also pointed out that an ambitious, profit-seeking Hydro One would logically want to sell power stateside and spread its new wings into other businesses rather than keeping prices low in Ontario. Despite the break-it-up impetus behind the whole Hydro restructuring, hadn't Hydro One already spent $523 million in buying up a large number of Ontario's municipal utilities?

But it was not enough to persuade Harris not to go ahead. Hydro One management, a group that included the premier's former close advisor Deb Hutton and was headed by the strong-willed Eleanor Clitheroe, was dead set on building a global company and did not want to hear any of this non-profit talk. (Hutton even upbraided a reporter who referred to Hydro One as a Crown corporation, explaining testily that it was a business with one shareholder.) Hydro One chair Sir Graham Day, who had earned his knighthood for his role in privatization in the United Kingdom, threatened to quit if the company was not privatized. TransAlta, which had set up an electricity retailer and a natural-gas-fuelled generator in Ontario, threatened a capital strike if the government did not follow through with full-blown privatization. The privatization express may have taken six years to get up a full head of steam, but it finally seemed to be on the move.

"I don't see any alternative," AMPCO president Arthur Dickenson said in December 2001. "If we stop now, what will happen?"

It was one of those bitter February evenings when the windblown streets of Canadian cities are deserted. But sixty people still showed up at the church hall behind Kingston's St. George's Cathedral. It was one of dozens of similar meetings held in the winter of 2002 across Ontario. The bar graph beside a portable pulpit showed a little bump dwarfed by two much bigger spikes. The price of a kilowatt-hour of electricity was 8.5 cents in Toronto, 16.9 cents in Chicago, and 35.28 cents in New York.

First up was Paul Kahnert of the Ontario Electricity Coalition. A stocky fellow wearing construction boots and holding bragging rights in Eastern Ontario, Kahnert had been among the Toronto Hydro linemen who had volunteered to chip ice from frozen lines during the great Ice Storm of 1998. He was anxious to pack everything he had to say into his ten-minute presentation, so he didn't mention the ice storm. Kahnert said he figured government had got a bad deal with the lease of the Bruce nuclear facility to British Energy, but he was mainly concerned about what would happen if Ontario succeeded in privatizing transmission and generation and started selling power to the Americans. NAFTA would kick in, along with the end of made-in-Ontario power policy. It seemed odd, but the twenty-three-year veteran of hands-on power-line troubleshooting and union activism was sounding a warning that was virtually the same as the one set out by Dofasco's CEO two months earlier: "It's absolutely impossible to have cheaper power prices when you add in profit."

Next up was Ontario NDP leader Howard Hampton, who left the pulpit and paced the church hall like a bible-thumping preacher, gesturing for emphasis, laying hands on the bar graph. "If you were a company operating a privatized generating station, would you sell power at this price?" he asked, pointing to the number for Toronto. "Or would you sell it here?" Chicago. "Or here?" New York.

The crowd was to some extent made up of NDPers and others likely to turn out for this sort of affair. But it also contained a good sprinkling of people who had not made up their minds, people who were wondering about all the talk about Enron and the comparisons between Ontario and California. John Gerretsen, Kingston's former mayor and sitting Liberal Member of the provincial legislature, had kept his ear to the ground and apparently decided that the issue was hot enough that he had better show up that night. He rose to confirm that his party, too, was opposed to the Tory electricity plans. This sparked hoots of disbelief from the partisans in the crowd, but Kahnert said he had a meeting set up with Liberal leader Dalton McGuinty for later that week and was still waiting on the Liberal position. The Liberals, who held a solid lead heading into an election year, were anxious to associate themselves with any sort of public discontent with the Tories, but they were having trouble staking out a clear position on the electricity issue.

Within days of the lively Kingston meeting, both Kahnert and Gerretsen confirmed that the Liberals, though opposed to the Hydro One sell-off, still supported plans to privatize generation and set up a competitive market in electrons. Buyers still seemed wary. Utilities Kingston reported that some 15 per cent of local households had signed fixed-price contracts with resellers but that most people were taking the path of least resistance and sticking with their public provider. If the public hadn't been paying much attention to the hydro file, that was changing. In Kingston, within a few weeks of the meeting at St. George's, the city council passed a resolution calling on the government to "cancel its plans to privatize and deregulate the electricity market." Eventually a total of forty-two municipal governments—from Galway-Cavendish and Harvey Township and New Liskeard to Toronto and Ottawa—followed suit, with resolutions opposing either the Hydro One privatization or deregulation or both. The Ontario Electricity Coalition was starting to tell itself that dressing its campaign up in a top hat and using the unlikely symbol of the capitalist who put an end to gouging by private power companies just might pay off. It had, at least at the start, seemed like a too-little/too-late effort because the privatization express had built up too much momentum.

The success of the anti-privatization campaign was evident as early as a Conservative leadership debate in February. Elizabeth Witmer surprised her opponents by calling for a rethink of the move towards electricity privatization. "As I have traveled this province . . . I continue to hear grave concerns about what is going to happen with the privatization of hydro," she said. "There is a lot of concern out there that the prices are going to skyrocket. Somehow we have to provide certainty and we have to better communicate." The comment drew audible groans from her opponents, who united to attack her for merely suggesting such a thing. A grouchy Ernie Eves pointed out, "We made this decision in cabinet three years ago." Former Energy Minister Chris Stockwell told Witmer that he was "dumbfounded" about how "suddenly it's dawned on you that maybe this wasn't a good idea."

The Ontario Electricity Coalition had its start when Charlene Mueller and Paul Kahnert decided that if no one else was going to organize against the wholesale restructuring of Ontario's public power system, they would. The two Toronto Hydro workers put two years of volunteer labour into organizing opposition to what they viewed as a threat to electricity workers and the public at large. They were not professional lobbyists, nor did they have any contacts in the corridors of power. They had no lawyers, at least at first. What they did have was the backing of their union local, an organization with attitudes and traditions much different from those of the union representing workers at the provincially owned power companies.

Kahnert, an avid martial arts practitioner and motorcyclist, became active in union affairs in 1983, when his co-worker Neil Morrison fell off a sixty-five-foot pole. The other crew members tried to keep Morrison alive with CPR, but he was too badly mangled. The next day, when Toronto Hydro wanted the crew to return to work, the men were not impressed. Electrical utility workers tend to be a militant group, proud of their skills, well-organized and reluctant to kow-tow to orders they regard as foolish or arbitrary. "We'd seen the most horrible thing you can imagine happen to a good friend and they just didn't give a shit," Kahnert said. "It was like a chicken had died."

The path to activism was different for Mueller. As a member of Local 1 of CUPE she had attended only a handful of meetings before finding herself, in 1988, in a non-traditional job. There she was, working as a inspector, a woman wearing the same green overalls as the male trades workers. At the time the local was wrestling with pay equity, and one of its leaders was a deft organizer and talent-spotter named Rob Fairley. He asked Mueller to be on the committee, and she slowly immersed herself in union work. She described herself as a Silent Sam who took her time, listening and learning about the complexities of union politics.

The Toronto Hydro workers did not see themselves as "production associates" or members of one big company team. They had responded sceptically to the alphabet soup of management team-building efforts like Total Quality Management and Quality Circles that came at them in the 1990s. They were different from their opposite numbers at CUPE 1000, the Power Workers' Union, though their ranks were also thinning. When Toronto Hydro merged with other municipal utilities, their numbers declined from 2,300 to 1,300. Still, CUPE Local 1 members voted to go on strike four times between 1989 and 1999. The 1999 strike was sparked by Toronto Hydro's argument that the local was negotiating its first contract because it was a new union representing not just workers in the former Toronto local but also workers in the former boroughs of the newly amalgamated city. According to this logic, all the hard-won worker-friendly language of the contract with the former Toronto Hydro local could be scrapped. "They wanted to start with a blank sheet of paper," Mueller said, recalling their victory in the bitter three-week strike that followed. "But we have a long history of 'We'll fight. No matter what.'"

By the time the hydro privatization issue came to a head in 2001, the Toronto Hydro workers had also learned about running campaigns. When management attempted to move most of their work from days to nights, they campaigned around the slogan "We are not night creatures." When the issue of contracting out the replacement of Toronto's streetlights came up, the theme of the video and the buttons was "We Light the Streets." Through it all, CUPE Local 1 activists became familiar with lobbying politicians and appealing to the public at large, picking up on the notion of marketing an issue through catchy slogans.

In 2000 Mueller and Kahnert approached Local 1 president Bruno Silano with the idea of doing something, anything, about electricity privatization. "There was no opposition in the province to the government's policy," Mueller recalled. "That's what sparked the fire under our asses."

Silano, relatively new to the job, was nonetheless aware of the success that the Power Workers had enjoyed by using the symbol of Niagara Falls during their big campaign against privatization in 1995–96. But it was still not an easy sell at the CUPE Local 1 executive board. It was one thing to organize around problems that had such a direct impact on the membership of one local, and quite another to go after a major issue of public policy. Kahnert remembered that it had even been difficult to get the board to give a hundred dollars to striking day-care workers: and here he and Mueller were asking for $50,000. They argued that Hydro privatization represented a broad threat to workers, comparable to contracting out street-lighting or meter-reading.

At the executive board meeting Kahnert held up the collective agreement and asked, "What is our number one responsibility as a union?" The answer: "The negotiation and protection of this agreement. Our work life and our home life depend on it. This is the biggest threat to come along in a hundred years." To his surprise, the board agreed to give them the $50,000—a huge sum for a local of 1,300 people.

In the organizing that followed, Kahnert became the public voice of the OEC as the organization gathered support from other union bodies, environmentalists, Canadian nationalists, and NDPers. Mueller remained in the background, marshalling information, running the website, and organizing the details. On one level, it was a typical division of labour. The more talkative Kahnert, who had already learned to say "power line persons" instead of "linemen," began to catch himself and apologize if he interrupted when Mueller was speaking. When Mueller did speak up, it was with understated passion. Later she remembered how, when making the case to get money from the local, for the first time in her life she found herself saying things like "We are going to win this. Here's why and that's why we need your support." She would argue that it was "the right thing to do. We're being robbed."

Rob Fairley, by this time an independent consultant, wrote a pamphlet arguing that it was not too late to stop the plans for deregulation, that deregulation would not only be unfair to people on fixed incomes but also produce economic instability and more pollution.

He pointed out that twenty-two U.S. states had shelved deregulation plans following the California debacle. But his main pitch centred on the issue that would have the most political traction. The cartoons by radical pamphleteer Tony Biddle showed bug-eyed people in homes, factories, offices, and hospitals staring slack-jawed in surprise at pieces of paper labelled "hydro bill." The oEC distributed half a million of the pamphlets starting in January 2002. By that time the Coalition was receiving financial support from CUPE's national office.

Acting on the advice of CUPE colleagues who were veterans of the long struggle to preserve public-health care, the oEC decided to take its campaign to communities where it was easier to get media attention. Kahnert spent the winter on the road, addressing gatherings organized by local chapters of the Council of Canadians, labour councils, and NDP riding associations. He spoke to fifty-five meetings in the four months leading up to market-opening. Kahnert and Mueller also convinced environmentalists at Greenpeace and the Toronto Environmental Alliance to come out publicly against electricity privatization. The green position of conservation and renewables gave the oEC a response to the inevitable question, "So if you are against privatization, then what are you for?" Kahnert sounded like Amory Lovins or Ralph Torrie. "Conservation is the fastest, cheapest and cleanest way to deal with our electricity crisis," he said enthusiastically. "But private power companies have no interest in selling less product. They've destroyed conservation programs wherever deregulation has taken root."

Meanwhile, Howard Hampton had sniffed the winds in search of an issue to animate his party's flagging fortunes. Popular disenchantment with the Tory government had been attaching itself to Dalton McGuinty's Liberals, and NDP tacticians figured the Liberals had a wishy-washy policy on the Hydro issue. By staking out a clear position in opposition to privatized power in a free market, they would benefit from any public wrath should the Tory policy go wrong. If that happened the Liberals, who had unanimously supported the market-opening enabling legislation and were cozying up to the Bay Street privatization industry, would scramble to denounce the Tories. At that point the NDP could occupy the high ground by denouncing the Liberals as opportunistic wafflers. Besides, Hampton brought a sense of personal mission to the issue. As a university student he had worked on an Ontario Hydro construction crew in Northwestern Ontario. He really did believe passionately in Public Power. By 2002 the NDP had painted up a campaign-style bus with those two words emblazoned on its side, together with pictures of Hampton and various electric appliances. The party even had a website, "publicpower.ca." Despite rumblings of internal discontent that the party was investing too much political capital into one issue, Hampton took his show on the road.

The NDP also commissioned an opinion from Steve Shrybman, one of the country's handful of left-wing lawyers who specialized in trade law. The party wanted to know whether privatization and open markets would leave Ontario's electricity supply "open for business" by companies wanting to export power to the energy-hungry United States. Shrybman knew about the way free trade allowed corporations to play fast and loose with the ability of governments to safeguard the environment. Like many of those who had bolstered the anti-nuke movement during its activist phase in the 1970s, he was a back-to-the-lander who still tried to spend as much time as he could at his Eastern Ontario farmstead. He had an interest in electricity dating back to the mid-1980s, when he had pushed the idea of an Efficiency Vermont-style energy conservation utility for Ontario.

The free-trade deal with the United States had given the U.S. negotiators what they wanted—unrestricted access to Canadian energy, particularly oil and gas—and Shrybman's opinion confirmed what the NDP suspected. The same principles of access also applied, at least in theory, to electricity. But in the case of electricity three roadblocks stood in the way: the existence of provincial public-sector monopolies, vertically integrated structures dominating generation, transmission, and distribution; the lack of the massive power lines needed to export large volumes of Canadian power; and various exceptions that shielded some provincial and municipal regulations from free-trade requirements. The NDP learned that the privatization enterprise would work to remove these roadblocks.

Alberta was the first jurisdiction in Canada to move towards a deregulated electricity sector. Alberta had always been the odd man out within Confederation when it came to electricity, as with so many other things. There had never been an Alberta Hydro, no provincially owned monopoly to privatize. Rather there had always been a mixed system in which private generators sold to municipally owned distributors in a regulated system similar to that in many U.S. states. As well, given the closeness of Alberta's dominant ethos—a firm belief in the benefits of freewheeling capitalism—to that of our southern neighbour, it came as no surprise to academic observers of electricity policy that Alberta was the first jurisdiction in Canada to move towards U.S.-style deregulation.

The province began the restructuring process as early as 1993, and had the legal framework in place by 1998. But it was caught in the backwash from California, and things did not go well. In the aptly named report ZAP! Alberta Is Jolted by Electric Deregulation, RBC Dominion Securities noted, "In unnerving similarity to California's experience, this process has resulted in Alberta's power prices increasing from amongst the lowest in the world to among the highest prices in North America."

The California reference was pertinent. B.C. Hydro opened its transmission lines to third parties in 1997 as part of a requirement it had to fulfil in order to obtain a U.S. Federal Energy Regulatory Commission licence. When it did that, because B.C. hydro lines are connected to Alberta's, it also connected Alberta to the U.S. market for the first time. When California opened its doors to competition in 1998, Albertans found themselves bidding against California for power—and paying Californian prices—even though they had ample supply within the province to meet their own needs. The province's prices rose from 1.5 cents per kilowatt-hour in 1996 to an average of 22.4 cents per kilowatt-hour in the fourth quarter of 2000.

The impact was muted by the ability of Premier Ralph Klein's government to dip into its oil-revenue wealth and send every Albertan a cheque to ease the pain. Going into an election, Klein announced a billion dollars in electricity rebates for 2001—over $320 per Albertan—and a rate cap that would be in effect until January 2002. Yet even after a landslide election victory, the Klein government still faced electricity headaches, with a price of 6.5 cents per kilowatt-hour in 2003. Meanwhile, the architect of the province's deregulation plan, Steve West, was musing that it could take until 2012 for the benefits of deregulation to kick in. Klein, ever the pragmatist, postponed the next stage of deregulation until 2006.

Although people with a rudimentary understanding of the new free-trade regimes believed that those regimes only began to bite in when international transactions in particular goods and services started, Shrybman's analysis explained that once companies with global ambitions (the U.S.-based Reliant Resources or Britain's National Grid Transco, for instance) got involved in Ontario's electricity market, formerly domestic matters of regulation and contract law would suddenly become business relationships governed by the powerful, business-friendly panels at NAFTA and the World Trade Organization's General Agreement on Trade in Services (GATS). If, for instance, Utilities Kingston started to buy power from a U.S. company with a new gas-fired generator in Ontario, or if it contracted work to a British-owned energy services multinational, it would be doing a foreign-investment deal subject to international trade rules, not to made-in-Ontario laws debated and passed by elected representatives. Attracting foreign-owned firms to Ontario was a clear objective of the Tory privatization enterprise. Shrybman found that an understanding of the impact of obscure trade disciplines on Ontario's electricity sector was in short supply among politicians and officials, and even among the business interests that were driving the privatization enterprise.

Lest his readers doubt the word of a former hippie, Shrybman tossed in the opinion of a man who he said "must be taken as an authority on the subject." Renato Ruggiero, the first director general of the WTO, had been frank:

> The right of establishment and the obligation to treat foreign service suppliers fairly and objectively in all relevant areas of domestic regulation extend the reach of the Agreement into areas never before recognized as trade policy. I suspect that neither governments nor industries have yet appreciated the full scope of these guarantees or the full value of existing commitments.

All of which stuck in Shrybman's craw. When Harris announced that Hydro One was up for sale, Shrybman got even more agitated: "It offended me quite profoundly that they thought they could get away with this without a fight." But when he began shopping the idea of a legal challenge around to organizations with the cash to go to court, no one seemed interested. The Ontario Electricity Coalition had previously asked him about the viability of a class action appeal against market-opening, but was now busy with its grassroots campaign. Shrybman said that the higher-ups at the Council of Canadians, whose local chapters were helping with the OEC tour, "listened politely."

The idea seemed to be going nowhere when Kahnert addressed a crowded meeting in an Ottawa church hall in March 2002. By that time the campaign was building momentum, and an OEC "train-the-trainer" meeting in Toronto had attracted a hundred people instead of the thirty that its organizers had anticipated. Although Kahnert had been led to believe that it was tough to get the big-city press interested in an issue being pushed by a trade unionist, he was surprised that the reporters and cameras turned out in Ottawa that night. He was equally intrigued that a fellow he had never met had a lot of well-informed questions. It turned out to be André Foucault, the number two official at the Communications, Energy and Paperworkers union.

Though impressed by the turnout, Foucault was not optimistic about the Ottawa meeting. The crowd was picking up on Kahnert's militant anti-privatization message, but there seemed to be no clear political way forward. A pragmatic trade unionist who had served as president of the Ontario NDP, Foucault knew all about Howard Hampton's power bus. As secretary-treasurer of a big union, he was also used to writing big cheques to lawyers. When Foucault took his impressions back to the CEP's Ottawa headquarters, he was surprised to find that someone else in the office was thinking along the same lines. Fred Wilson reported that he had just heard Steve Shrybman's pitch.

"I thought I'd harass Fred," recalled Shrybman, who had met the CEP staffer when he was working for the B.C. Environmental Law Association. "He was one of the few people I hadn't harassed."

Wilson, an assistant to CEP president Brian Payne, had made common cause with environmentalists on the forestry and energy issues. Between them Payne, Foucault, and Wilson agreed to underwrite a court challenge to the Hydro One privatization, due to go ahead in a matter of weeks. It didn't matter that they represented no electricity workers; they saw the privatization enterprise as a matter of the overriding public good, transcending the self-interest of individual workers or union locals. CUPE, which did count power workers among its members, was their natural ally. But for CUPE national president Judy Darcy and staffer Morna Ballantyne, the women who had been making the decisions about helping the Ontario Electricity Coalition, the internal union politics were thornier than they were at the CEP.

One of their biggest and richest locals happened to be the unit representing Hydro One workers. Local 1000 had always forged a separate identity within CUPE, even branding itself—logo and all—as the Power Workers' Union. What's more, it had decided to get onside with Ontario's electricity policy despite the labour movement's rock-ribbed opposition to privatization. The PWU was aware that selling off public-sector assets invariably eroded wages and job security. It had previously attacked Britain's privatization pioneers, one of whose explicit goals was reducing union power; British firms had succeeded in eliminating some 66,000 electricity industry jobs by 1998. Still, Tory Energy Minister Jim Wilson had appeared at a union meeting and on the cover of the Power Workers' magazine. John Murphy, the local's president, served on the government's Electricity Transition Committee.

All of that activity caused a domestic dispute within the House of Labour, and feelings became even more bitter when Murphy, the architect of the CUPE Local 1000 flip-flop, went over to the other side. By the time of the court challenge to Hydro One privatization, Murphy was sitting in a vice-president's chair at Ontario Power Generation. Although no one in a position of authority at CUPE was willing to comment officially, many would have agreed with the assessment of Jim Stanford, the high-profile economist for the Canadian Auto Workers: "There is nothing new about being a management stooge."

Factory managers who are shrewd enough to spot talented shop stewards headed for union leadership positions often hire them as supervisors. "You make 'em, we take 'em," is the boss's approach. Those who change sides are particularly valuable to management. They make the best foremen because they understand the union culture and exhibit the predictable zeal of the freshly converted. So it was with John Murphy; he was soon talking the language of partnerships and pragmatism. "If you keep saying you are opposed to change but don't have any alternatives, you are going to get marginalized, and are not going to have any opportunity for influence," he said after defecting to OPG. Speaking of the Harris regime, he said, "We have not experienced any government that has done such broad consultation with the union on every step of the electricity industry restructuring."

The Power Workers remained choirboys for the privatization enterprise even after their leader decamped. It was unclear as to what kind of deal had been arranged with the government in exchange for the union's support. What was clear was that, aside from the police unions who loved the government's toss-em-into-boot-camps line on crime, no other labour group in the province was as close to the Tories as was the Power Workers' Union. Yet despite it all, Judy Darcy and Morna Ballantyne were still surprised when CUPE 1000—one of their own locals—retained counsel to intervene against its parent body after CUPE national agreed to go to court with the CEP.

Once the unions retained Steve Shrybman and his colleagues from Sack, Goldblatt, Mitchell, the lawyers had to rush to file the papers that spring, before the Hydro One IPO went through. The prospectus was hinting at a feisty company coming to market with bold expansion plans. It was already planning $590 million in capital spending. "We may pursue acquisitions of other transmission and distribution assets in North America, which could require substantial amounts of financing," said the prospectus. That could well have sounded off-putting to cautious investors who had been reading about Enron and were hoping for the solid, reliable dividends of a conservative transmission utility. But the financiers busying themselves on the deal, expected to close in May, were aware of the CEO's grand plans. "Eleanor Clitheroe," said one investment banker, "has always wanted to position this as a growth story."

No one on the union side was convinced that they would win in court. It was, at best, a long shot. Their argument was that governments that had privatized major Crown corporations like CN Rail or Air Canada had passed legislation granting themselves the authority to make such a major public policy move. Against this background, the unions argued that the legislation that divided the former Ontario Hydro into separate companies did not explicitly give the government the power to turn around and sell them. For its part, the government trotted out the old claim that unions should mind their own business with respect to anything that doesn't directly involve their members. Its main case, however, rested on the argument that as the owner of Hydro One's shares it could do whatever it wanted with them. The Power Workers predictably avoided their ally's first argument, but backed the second one.

By the time Justice Arthur Gans had heard the competing claims and was ready to deliver his judgment, Eves had won the Tory leadership and been sworn in as premier. The contending parties gathered on Friday, April 19, 2002, at a courthouse on the broad, flowered boulevard of University Avenue, just up the street from the spot where the Ontario Electricity Coalition had launched its anti-privatization campaign four months earlier at the Adam Beck memorial. As soon as Justice Gans started to read his decision to the crowded courtroom, it was obvious that something dramatic was in the wind. "Right away there was this buzz that went through the whole courtroom," said Cecil Makowski, the CEP's Ontario vice-president.

"'Hydro One,' the corporate name for the new millennium, is one of the amoebic offspring of Ontario Hydro created by the Government in 1998," Justice Gans began. "Ontario Hydro was one of the defining characteristics of the Province, one with which its residents could identify. . . . Its creation and basic foundation was the primary reason a knighthood was bestowed upon Sir Adam Beck in 1914. His sculpted image stands watch over University Avenue."

The judge went on to brush aside the argument that the unions were being "mere busybodies or officious meddlers" in bringing the case. He then traced the origins of the government's policy of introducing competition into the electricity sector, showing that when it passed the Act it did not set out as one of its purposes the privatization of any of its assets. Nor had its white paper included anything more than an oblique reference to privatization. The judge even quoted the energy minister, in introducing the Electricity Act, as ruling out privatization. "Privatization of a long-standing important public institution, such as Ontario Hydro," Justice Gans stated, "is not something I would have thought would or should occur without addressing the issue head on."

Justice Gans pointed to the glaring hypocrisy reflected in the government's policy. The Tories had been advertising the Hydro One privatization as a means of raising capital to pay down Ontario Hydro's leftover debt (blaming the debt on previous administrations, which they habitually painted as reckless spenders), and the prospectus had just indicated that the proceeds of the IPO would be used for that purpose. But there was no reference to debt repayment in the government's sweeping Electricity Act of 1998 and the government was obliged to pay the proceeds of any sale into its Consolidated Revenue Fund. As the buzz in the courtroom intensified, Justice Gans described this factor as an "apparent lacuna" in the law, adding that it stood in "marked contrast" to the explicit obligations thrust upon municipalities that sold off their utilities. By law, the money that the municipalities received had to be used to pay down utility debt.

Either the Harris government had not intended to sell off Hydro One four years previously, or it had blundered by writing a law that was not worded in such a way as to allow it to privatize the cash-cow utility. Whether or not the Eves Tories would have kept at least some of the proceeds in general revenues, using them to erase the deficit that they knew would be facing them going into the next election, will probably never be known. But the temptation would certainly have been there. After all, this was a government that traded on its reputation for fiscal probity. It was also inclined to sell assets rather than bring in more cuts to health care and education before an election.

Justice Gans's decision came down in favour of the unions, stating that the province did not have the authority to carry out the sale. The result was, of course, greeted with euphoria among anti-privatization campaigners, prompting them to keep up the pressure and redouble their local efforts to persuade municipal councils to pass motions opposing the government's electricity policy. The NDP, which saw it as their issue, peppered the new government with embarrassing questions. The Liberals, sniffing blood, also began to focus on the Hydro issue.

If the opposition was emboldened, the supporters of privatization were aghast. Apparently unaware of the vanguard role that Chile's dictatorship had played in electricity privatization, one New York investor complained that the ruling was "the kind of thing that happens in Latin America." Bay Street, having been enticed by rumours of a further bonanza on a possible $8-billion privatization of OPG, was outraged to learn that its hundred-million-dollar payday for the Hydro One deal had been, at best, postponed. Their trumpets in the business press reacted with florid fulminations about an "ideological cabal of big unions." *The National Post*'s Terence Corcoran warned that the Hydro One "disaster" was "the first of many tests that will test the soggy wet mettle of the new Eves government." Quoting Power Workers' Union president Don MacKinnon as saying (mistakenly) that Hydro One was broke and that "the private sector is the way to go," Corcoran declared, "That's a union voice the new government should listen to."

Premier Eves had been forced to listen to some other voices while campaigning that spring in and around Orangeville, where he had won a by-election. Apparently Elizabeth Witmer's political instinct had been correct when she warned the party about a growing public unease with its Hydro privatization plans. The voters in the sprawling new suburbs around Orangeville and on the back concessions seemed increasingly anxious about what was going to happen to their power bills and their public power company.

Even before the court decision, there had been rumblings in the Tory ranks. Rumour had it that, unlike Harris, Eves was not willing to make the tough choices. The backers of the privatization enterprise knew that the market was due to open in ten days' time and that the Hydro One privatization was simply stalled. It could easily proceed within weeks if the Conservatives passed a quick bill empowering the government to go ahead with the IPO. But something had happened. As they say in the world of sports, there was a sudden shift in momentum.

In the following six months Ontario's privatization enterprise was walloped by a series of body blows that left it staggering. At first the much-delayed May Day market-opening went smoothly enough. There were no price hikes. In fact, power prices declined slightly and the public took little notice. People in the utility business were certainly fascinated by the novelty of a fluctuating market price for electricity. But even they were distracted by the news that Premier Eves had sniffed the political wind. At the beginning of May he announced that the Hydro One sale was off the table. He declared that he would not be pressured by "some guy on Bay Street"—which sent the guys on Bay Street reeling.

No sooner had that news sunk in than the government turned on Hydro One for behaving like a private-sector company. By the end of May something the government had long known about had become common knowledge. Top Hydro One managers were preparing themselves to play with the big boys in the United States, and their pay packets had begun to reflect their ambitious plans. Within two years CEO Clitheroe's "compensation," as they call your paycheque at that level, had more than doubled, to $2.2 million. She had a car allowance of $174,000 and had been paid $172,000 for a vacation. Hydro One had spent $360,000 sponsoring a yacht—and Clitheroe was a known yachting enthusiast. Clitheroe, who pointed out that Hydro One had taken on many "private-sector responsibilities," also had a golden parachute deal, recently amended by the company's board. Her original severance package would probably have seemed more than sufficient to most citizens: it had been worth some $6 million and included a hefty lifetime pension. But just after the Gans decision and the government's wavering, the board changed the way the deal would be triggered, so that the CEO would get the money if there were "a fundamental change in the policies of the province relating to Hydro One." According to the vocabulary in vogue, the euphemism for this sort of extravagance was that Hydro One faced "governance challenges."

The Eves government, aware of how bad this sort of thing smelled in the Enron era, figured that it would look good if it came down hard on corporate greed. It dismissed the Hydro One board in early June. Soon after that, when Eves confirmed that the Tories were not going to sell the company, he repeated his new-found populist line: "I'm not here to please some banker from New York."

The Tory attempt at damage control sent the marketeers into a tizzy. "What an awful situation," complained Energy Probe's Tom Adams, sounding a quaintly naive note. "Now we've got politics thoroughly at work in our transmission system." Hydro One board member Bernard Syron, dismissed in the fracas, appeared to be somewhat confused at the turn of events. "I don't know what has happened to

Ernie. He blinked over the IPO. He isn't half the man [that] his predecessor was." (Syron had formerly been the boss at Cara Operations—
Swiss Chalet, Harvey's—where the board had underlined the private
sector's leanness and efficiency by topping up his annual $400,000
pension with a $615,000 bonus in 2001.)

Soon enough it would be Clitheroe's turn to follow her board onto
the tumbrel and head off for the Tory guillotine. The cynical political
calculus was that firing the CEO ("Heads will roll!") and the generalized outrage that would surely greet her perks would somehow distract the public from wider issues. It was revealed that, as the mother
of two recently adopted babies, she needed a limousine service that
allowed her both to work on her way to the office and to ferry her children and nanny around. The cost was $330,000. Those inhabiting
the lofty levels of corporate power were apparently living in another
world. It was certainly a world apart from pregnant welfare mothers.
The government was still trying to deal with the fallout from the
death of one such woman. She had been stuck in her stifling attic
apartment during a heat wave while under house arrest for not telling
the welfare police she had been studying while collecting benefits.

Firing Clitheroe did nothing to remove Hydro affairs from the
public spotlight. Hydro continued to dominate Question Period at
Queen's Park to an extent not seen since the late 1970s, when the
corporation's free-spending ways had given rise to a royal commission and a special legislative inquiry. The opposition grilled the government zealously. Then, on June 11, an early summer heat wave sent
the temperature past 30 degrees Celsius in smog-choked Southern
Ontario, thrusting market speculators into overdrive. The resulting
price spike was brief, but it was a taste of things to come. Any time
the thermometer topped 30 degrees Ontario would have to import
power because of the shrinking cushion between supply and demand.
Dave Goulding, CEO at the new IMO, reported that Ontario power
plants had a capacity of 27,000 megawatts the previous year, when
peak summer demand was 25,000 megawatts.

Plagued by uncertainties on the supply side, the system was
obviously operating with a dangerously slim margin of error. The
moribund nuclear reactors at Pickering A and Bruce A were still
sucking up cash but seemed no closer to producing electricity. The

private companies that had applied to build new gas-fired plants had yet to break ground because of uncertainties over both the market price of power and the manliness of the provincial government. On the demand side, the Conservatives had a do-nothing, let-the-market-decide attitude that had been succinctly summarized by former Energy Minister Jim Wilson. The previous summer Wilson had been asked about bringing in conservation measures as an alternative to relying strictly on new gas plants that still polluted the environment. Wilson replied, "The private sector asked us to get out of large-scale government conservation programs." Morever, he added that previous conservation efforts "may have made the odd person feel good, but they had absolutely no effect."

The campaign against privatization was having its effects behind the scene. The May 2000 moratorium on selling off coal plants had been lifted sixteen months later, after a government review concluded that privatization would not result in damage to the environment; but by that time the terrain had begun to shift. Both Greenpeace and the Ontario Electricity Coalition were regularly denouncing potential sales as a licence to pollute, and the Ontario Clean Air Alliance was demanding that any sale be conditional on converting the coal burners to natural gas. When Ontario Power Generation put the Atikokan and Thunder Bay coal plants on the block in late 2001, it did find a buyer. However, when OPG asked the provincial government (which, after all, had passed the legislation requiring it to sell the generators) to ratify the sale, the cabinet refused for "environmental reasons." The government had changed its mind and was now demanding that the plants be converted to gas, something that the buyer was not prepared to do. According to OPG vice-president Richard DiCerni, it was "like selling your house when your father holds the mortgage. You enter into an agreement and then your father says, 'Well, I'm not selling unless the person builds a swimming pool.' Then their guy says, 'That's not the deal.'"

With the sale of the coal plants off the table due to concerns over smog and climate change, and the water-driven turbines off limits because they were part of Ontario's cultural makeup, selling or leasing the nukes appeared to be the best option. Yet in the middle of the fuss over Hydro One and the hot weather, the provincial auditor, Eric

Peters, released a report shedding some light on the extent to which Ontario's citizens had been fleeced when Ontario Power Generation leased the Bruce nuclear station—at full capacity capable of supplying a quarter of Ontario's power needs—to British Energy. Or at least that was the part of the report that, understandably, received the most public scrutiny. In a little-noticed section the auditor echoed Justice Gans's point about government treatment of the proceeds from the disposal of Crown electricity assets. The Harris government had received an initial payment of $370 million as part of the deal with British Energy. The money was allegedly to be used for paying down the Hydro debt. If so, it should have gone directly to the Ontario Electricity Financial Corp. Instead, the government kept the money in general revenues, where it could do with it as it pleased: perhaps another pre-election tax cut. Maybe the Tories intended to use it to balance a budget that, despite years of steady economic growth, was close to a deficit because of previous tax cuts. The auditor, a cautious man, had no way of knowing and was certainly not in the business of speculating on political motivations. His job was to report the facts.

His main finding was that the lease deal with British Energy had resulted in $214 million in lost profits for Ontario Power Generation. The province would also see a cash-flow reduction of $170 million, mainly due to the payment of federal taxes by the private operator. Had OPG kept running the reactors, the money would have remained in the company's coffers because it did not pay federal taxes. Peters could not speculate about how many millions in lower profits the province would forego in the future, but the report indicated that the lease, signed to inject competition into a newly opened market otherwise dominated by OPG, was a bad deal—and it was especially bad compared to similar transactions in the United States, where nuclear assets changing hands had been valued at twice the price that Ontario received from British Energy. In the U.S. deals, the immense liabilities unique to the nuclear power industry were transferred to the companies purchasing reactors. Because the government had only leased the Bruce facility, some $5 billion in liabilities for reactor decommissioning and disposal of poisonous waste stayed with the public— while the private operator pocketed profits that promised to be enormous with market-opening. Should prices rise merely by 10 per cent, Peters pointed out, the public operation of the Bruce reactors would net the government $565 million. As it stood with the sweetheart lease deal, the government would simply get an extra $280 million.

Defenders of the deal could point to the huge amounts that Bruce Power was reinvesting to get the aging Bruce A reactors going again. There was also the high cost of maintaining any Candus—including the working reactors at Bruce B—as they approached the end of their design lives. These big, expensive jobs entailed risk, which meant that the company was entitled to a good return on investment. All of which, of course, made for powerful arguments for letting the nukes run down and die a natural death, and for setting aside the surplus proceeds from their continuing sales to deal with radioactive garbage and invest in conservation measures and renewable energy.

Critics of Ontario's nuclear power experiment had been wondering for five years about the ill-fated Pickering A, still undergoing a refit. How much money was going into that sinkhole for capital? The auditor was wondering the same thing. Peters was worried about the cost overruns and tardiness in repairing the four aging Candus at Pickering. No one outside OPG knew what it was costing, but Peters mentioned a billion dollars in overruns, plus delays that would keep supply down and drive prices up.

By this time OPG and Bruce Power executives, along with municipal utility managers and industrial power-users, were logging onto the IMO website to watch the price of electricity fluctuate on the spot market. One in five households, or about 850,000 consumers, had signed contracts with private-sector resellers, who were equally keen observers of the daily and even hourly ups and downs of power prices even though they were likely to have taken out firm supply contracts with OPG. The unexpected hot weather in early June 2002 and a sudden spike in price—$700 per megawatt-hour during a day when the average price was $76—were foreboding signs. So was secrecy. Soon after market-opening, Ontario's Information and Privacy Commissioner protested the secrecy rules with which the government was shrouding the IMO. Anne Cavoukian insisted that with the open market the public interest could only be served if the IMO's operations were "open and transparent." That way, California-style market manipulation could be more easily avoided. But the government, which had already exempted Hydro's successor companies from the Freedom of Information laws, responded that the new entities could not disclose sensitive private-sector information because doing so would dissuade companies from investing in new generating capacity. In fact, private power operators were so edgy about the uncertainties in Ontario that they would rather have eaten glass than committed themselves to building new generators in the province.

⊕

The summer of 2002 would go down as one of the hottest on record. The Canada Day weekend saw temperatures rise into the mid-thirties in Southern Ontario, with the IMO issuing its first "power warning." In Southern Quebec and Southeastern Ontario the midday sun turned into a faded yellow ball as out-of-control forest fires in Northern Quebec filled the skies with wood smoke. Electricity prices leaped to 4.7 cents per kilowatt-hour in early July compared to their 3.2 cents average in the two months following market-opening. An open market seemed to be leading to wild price changes. By the end of the scorching month, temperatures had topped 30 degrees Celsius for sixteen con-secutive days, and the IMO was issuing regular warnings about the need to cut down on power consumption.

Weird weather was not confined to Ontario that summer. Western Canada faced another year of drought, as late-May forest fires raging in Saskatchewan and Alberta forced people from their homes and left them praying for rain. In mid-June the Prairies were hit by flash floods that washed out roads and flooded farms. As the Western premiers called for a national debate about climate change, Ottawa announced a cost-sharing agreement worth $5.2 billion to farmers struggling to cope with drought and floods. By mid-July governments were telling farmers that if they wanted to survive they would have to abandon some of their traditional crops and diversify into livestock in the face of uncertain weather. In August Statistics Canada warned that the harvest of 2002 would be one of the worst in Western Canadian his-tory, with Alberta farmers hit the hardest. Many Prairie producers could not even grow enough hay to feed their existing livestock. To help them out, Eastern farmers began to load donated hay onto trains as part of a Hay West campaign.

August also saw Central Europe's biggest relief effort since the Second World War as floods ravaged Germany, Austria, and the Czech Republic. Hundreds of thousands of people fled their homes as rivers overflowed, ravaging architectural treasures in Prague. The damage was estimated at $30 billion. That summer a report from Natural Resources Canada chronicled the horror show of environmental degradation that would result if temperatures continued to rise. Illness would spread from contaminated water as supplies of potable water shrank. Fish species would disappear when their spawning areas dried up. The severe drought on the Prairies would be more likely to continue, with predictably damaging effects on crop yields. Water levels in harbours and waterways—from the vital St. Lawrence Seaway to vacationland marinas—would shrink. The lower water volumes would create problems for the turbines at Ontario's huge dams at Niagara and on the St. Lawrence. The Natural Resources report predicted reduced water flows in these Southern regions, particularly during the hotter, drier summers, precisely the time when pressure on the electricity system was mounting.

Bizarre weather, characteristic of the chaos predicted by climatologists long concerned about the buildup of greenhouse gases, had become the norm. It was a vicious circle: burning fossil fuel to generate electricity was disrupting weather patterns and making things hotter, which in turn led to greater use of electricity for air conditioning while reducing water flows and the ability to generate hydroelectric power. Against this background Jean Chrétien's federal government decided that it was time to ratify the Kyoto Accord, setting off a firestorm of political opposition from the petroleum lobby and several provinces. In Alberta, a province being devastated by drought, Premier Ralph Klein, not known for his nationalist posturing, led the charge against the Accord, arguing instead for a "made-in-Canada" approach to curtailing greenhouse gas emissions. Ontario joined the opposition in the fall, raising the spectre of mass unemployment if Kyoto were implemented and arguing that every province should have its own solutions. The dust-up featured rival polls—Canadians support Kyoto/no they don't—commissioned by the opposing camps. Full-page newspaper ads from oil companies argued that there was no need for strict emission rules: guidelines and voluntary programs would suffice. Critics responded by pointing to the laissez-faire approach to auto emissions standards, which had given rise to the boom in wasteful, high-pollution vehicles. In the midst of it all Ottawa released data showing that Canada had produced 19.6 per cent more greenhouse gases in 2000 than it had in 1990. That finding was not entirely due to the spread of fuel-hogging SUVs, those great panzers of suburbia; nor did it only come about because unrestrained suburban sprawl demanded automobile use. Canada's relationship with the gas-guzzling economy south of the border, cemented by NAFTA, was a huge factor. In the course of the decade oil exports to the United States had jumped by 328 per cent, gas exports by 158 per cent. That activity more than doubled the greenhouse gas emissions resulting from petroleum production for export. This pollution rose from 28 megatonnes in 1990 to 65 megatonnes in 2000. As more U.S. electricity utilities switched to natural gas, the trend would only continue.

Despite its gravity, the Kyoto debate was made up of rather intangible elements. What was very tangible that summer was the air, particularly in Southern Ontario, where people could often see what they were breathing. The province had finally included fine particulates in its air quality index, and the result painted an even bleaker picture than the one portrayed by anti-smog activists, who had long pointed to the figure of one thousand premature deaths a year caused by bad air in Toronto alone. The summer of 2002 saw twenty-six days during which the air was so fouled that breathing was considered dangerous. The summer had twice as many days as usual in which the temperature topped 30 degrees Celsius, and the situation was worsened by a drought that produced the driest August on record. The hot weather, combined with a lack of generating capacity and a volatile open market, drove electricity prices higher. Power bills soared between 50 and 100 per cent above those in the previous year. In July the average wholesale price was up sharply, at 6.2 cents per kilowatt-hour, compared to the 4.3 cents at the May 1 market-opening. In September the price hit 8.3 cents. The price jolts on Ontario's new open power market were so dramatic that they even pushed the national inflation rate up by half a percentage point.

The down-but-never-out nuclear power industry glommed onto Ontario's environmental chaos, arguing that its reactors offered a clean, weatherproof alternative to other sources of electricity. The controversy over Kyoto ratification gave the industry another angle to exploit in its extensive advertising and lobbying campaign. "If Canada is to achieve its climate change and clean air goals, nuclear energy is absolutely essential," said Bill Clarke, president and CEO of the Canadian Nuclear Association. But the industry's claims were yet again tossed into doubt with the news that only one of the four reactors at Pickering A would be starting up in 2003, and at a cost of an additional $230 million. John Baird, Ontario's third energy minister in as many years, confirmed that the future of the other three reactors was still in doubt. He grumbled aloud about the project being four years late and $1.4 billion over budget. The uncertainty added to the confusion about the state of Ontario's electricity system, giving private power producers another reason not to invest in new power generation.

There was more to come: the fall of 2002 brought the news that British Energy was headed for bankruptcy. The company to which the government had entrusted the Bruce nuclear power station—the owner of 82.4 per cent of Bruce Power and operator of twenty-six nuclear reactors in the United Kingdom, United States, and Canada—had lost over a billion dollars in 2001. The crisis prompted Canada's nuclear regulator, the Canadian Nuclear Safety Commission (CNSC), to demand weekly updates from Bruce Power on the state of the financial guarantees from its parent company. These commitments included access to $222 million in shutdown money in the event of an emergency. Bruce Power boss Duncan Hawthorne, who had come to Canada from British Energy, suddenly found himself in a position familiar to managers of nuclear facilities. Taking the offensive, his report to the CNSC indicated that he was "insulted" by any musings that lack of money could undermine safety.

British Energy itself was a case study in the privatization enterprise at work. When the British government privatized power, it found no takers for its aging and problem-ridden Magnox nuclear plants and was forced to create a new state-owned company to run them. The rest of the nuclear plants were moved to a new company called British Energy, which was privatized in 1996 after the government agreed to underwrite the costs of decommissioning and created a legal obligation for electricity distributors to buy "non-fossil" electricity or pay a penalty, which in turn acted as a subsidy for nuclear and (to a lesser extent) renewable energy. When British Energy hit the wall, the government—unable to have the supplier of 20 per cent of the country's power go bankrupt and unwilling to abandon the nuclear power plants, with no organization in charge of decommissioning them—came up with the emergency cash. The nuclear bailout cost the British government over £3 billion; it had originally received £2.1 billion for these assets. The company's woes resulted in part from a decline in the price of power due to low fossil-fuel prices and overcapacity. British Energy, with nuclear power's higher costs of production, could not compete. By the time it went into crisis in 2002, it was the only one of the original electricity suppliers still controlled by British capital. The rest of the major players had been swallowed up by European multinationals, which were themselves conducting a wave of concentration, a "rebundling" of the market that involved consolidating generation and distribution assets. The market was increasingly dominated by a few powerful corporate hands: Electricité de France, E.ON and RWE of Germany, First Energy and MidAmerica of the United States.

As one of the conditions of the Blair government's bailout, British Energy was obliged to sell its 82 per cent share of Bruce to help pay down its debts, even though the favourable terms of the lease had provided Bruce Power with $106 million in profits in its first year. Taking over from British Energy in 2003 were the Saskatchewan-based uranium mining firm Cameco, TransCanada Pipelines, and the Ontario Municipal Employees Retirement Board (OMERS), which each bought 31.6 per cent of the shares. The Power Workers' Union and the Society of Energy Professionals each doubled their shares in the plant, to 4 per cent and 1.2 per cent respectively. Oddly enough, many OMERS contributors were also members of CUPE, which had vehemently opposed the whole privatization scheme. Like most unions, CUPE had no control over the pension plans it negotiated on behalf of its members.

The travails of a British nuclear power operator, however, were the least of the Ontario government's worries. The province's long-standing commitment to nuclear power, combined with Tory faith in the magic of the market, had made the government very nervous indeed about the state of its power system. By autumn 2002 eight of the province's twenty nuclear reactors had been out of service for five years—precisely the period during which the government's privatization enterprise had gained momentum. For the free-market experiment to have had any hope of succeeding in its advertised goal of keeping prices down—as late as 2002 the government was still insisting that competition would bring $6 billion in savings to electricity users—supply would have to exceed demand by a comfortable margin. For that goal to be reached, the nukes would have to pump more power into the grid. As well, more power would have to come from a new fleet of gas-fired generators that the private sector was supposed to be building. The IMO had fifty applications on file, but few plants had been built even though Ontario's reserve margins were amongst the slimmest in North America. Private operators were apparently as spooked as the government. How could they compete against OPG, which still dominated the market? Would all the money that OPG was dumping into refurbishing nuclear plants pay off, boosting supply? If so, when? How would that influence prices?

No one seemed to have the answers. As the fall shoulder season arrived, power prices remained stubbornly high. While the big industrial users were protected from the rigours of the market by their closed-door deal with the government, others were not so lucky. These included small-business operators and middle-class suburbanites who had hungry electricity habits—groups that also comprised the core Conservative constituency. The government's mind was further concentrated by the reality of hydro rates in Toronto, where the municipal utility had been shielding customers from spot-market increases, which had held steady since May; bills were set to soar when Toronto Hydro started to play catch-up. By the end of summer 2002 Toronto Hydro was owed $619 million and had been financing the unpaid bills through bank loans. The utility had just paid the IMO $276 million for power in a single month, up from its previous record of $154 million.

Tory backbenchers began to feel the heat from angry constituents. MPP Cam Jackson, who had just been booted from the cabinet over allegations about his personal expenses (of which he was subsequently exonerated), called publicly for a rate freeze. Veteran Tory Gary Carr, the Speaker of the legislature, and who hailed from the party's heartland in Oakville, had already demanded an end to the six-month fling with market power-pricing. This sort of internal dissent inevitably makes the opposition salivate, and both the Liberals and NDP redoubled their attacks on what was now routinely being described as a "botched" privatization or deregulation scheme—even though little had been privatized, and electricity had never been independently regulated in the first place. The die was finally cast when Tory labour minister Brad Clark broke cabinet solidarity and joined the call for a cap on electricity prices. On that very day, November 4, the price of power was 6.9 cents a kilowatt-hour in the late afternoon, up from the 5.17-cent average since market-opening—and this was during the period of moderate demand.

A week later, on November 11, Premier Eves travelled to an upscale neighbourhood of suburban Mississauga to visit a 3,500-square-foot suburban home. In the kitchen was a fridge with a built-in icemaker; in another room, a television with a fifty-inch screen. The place prompted one press corps wag to remark that the family living there didn't need lower hydro rates, it needed its own generator. But lower rates were what the Hardatts and everyone else in Ontario got. Staring into the teleprompter beside the giant refrigerator, the premier announced that for four years prices would be frozen at the pre-market-opening level of 4.3 cents per kilowatt-hour. Everyone would get a rebate cheque in time for Christmas.

If the Remembrance Day reversal seemed like a desperate, cynical political move, it was. On one level, it worked. Once the price freeze went into effect the hydro issue faded from Ontario's political radar screen. But on another level, the reversal damaged the image that Harris and Eves had been cultivating for seven years. They had painted themselves as being prudent managers capable of running things in a competent, responsible way—of delivering what they promised, even if was painful for voters. Now they were seen more as a desperate gang of bumblers. The price freeze would cost the public purse hundreds of millions of dollars because the market would continue to set the price of power. The government would be making up the balance between the 4.3-cent price cap and the market price. Not only that: because there would be no money forthcoming from the failed Hydro One privatization, the tax-cutting Tories—so proud of their balanced-budget legislation—would face an even deeper deficit.

Along the way the governing party had also alienated an important constituency. In the face of generalized public disquiet over the hydro file, the Eves government was forced to back away from the privatization enterprise and those who stood to gain so much from that enterprise. Veteran Queen's Park watcher Graham Murray, whose contacts extended from ministerial offices to Bay Street suites, reported that the rift was profound. "They had to break faith with their business community partners in a very jarring way," he said. As a result the business sector began "to say very rude things about not being able to trust the government: 'These are not the kind of people you can do business with.'"

All of these events had ominous implications for the government of the day, not to mention the privatization enterprise—and in the end Ontario was still left with the seemingly intractable problems of electricity supply. It did not have enough electricity ready to hand, yet its factories, office, stores, and homes used too much. Producing that power was already contributing to climate change and damaging the Earth in frightening ways. Scientists from Laval University and the University of Alaska who had travelled to Nunavut reported that a 3,000-year-old Arctic ice shelf had disintegrated over the previous two years. The breakup of the Ward Hunt Ice Shelf across Disraeli Fjord on Ellesmere Island corresponded with increases in warming since the 1960s. According to Laval polar ecologist Warwick Vincent, "a critical threshold" had been passed.

nine

Morbid Symptoms and a Regime Change

The energy system a society adopts creates the structures that under-lie personal expectations and assumptions about what is normal and possible. A child born into a world filled with automobiles takes them for granted and learns to see the world "naturally" at 60 miles per hour. . . . Each person lives within an envelope of such "natural" assumptions [that] together form the habitual perception of a sus-taining environment that is taken for granted as always already there.

— David Nye, *A Social History of American Energies*, 1998

THE VIEW HAS LONG attracted cottagers and campers to the lakeshore: startlingly blue water, breakers on the sand beach far below, turkey vultures looping lazily in the updraft as the breeze catches the steep cliff. At the Visitors' Centre at the top of the cliff, tourists are greeted by a mural depicting both an Aboriginal woman adorned in beads and feathers and a technician in a white lab coat, hard hat, and safety glasses: nature blending easily with modernity. Visitors learn that the area between the cliff and the lake is home to 235 species of plants and over 300 species of wildlife. It is also home to the world's second largest nuclear power establishment.

At the bottom of the cliff the Bruce Power site is its own city-state, a tribute to the heroic gigantism that is nuclear power. Aside from all the wonders of nature, the site has 56 kilometres of roads. It has its own firefighting and snowploughing crews, its own offices, restau-rants, laundries, training facilities, and warehouses. It also has the storage facility—more plainly, the dump—for all the low-level radioac-tive waste generated by Ontario's forty-year adventure with nuclear power. "Radiation," says a comforting sign at the Visitors' Centre, "is all around us." It is as "natural," the sign implies, as all those plant and animal species that are also all around us.

Only a few guests qualify for the trip down from the cliff to the former Ontario Hydro power station on the Lake Huron shore because tourist visits were halted after September 11, 2001. Those who do make it into the 3,100-megawatt Bruce B plant, which provides enough power to supply Toronto, cannot help but be impressed by the scale of the place. Hard-hatted workers use adult-sized pedal tricycles to shrink travel times in a turbine hall on the scale of the West Edmonton Mall. Duncan Hawthorne, the Scottish engineer who started running the place when Ontario Power Generation leased it to the private sector, wears a watch the size of an Oreo cookie.

Everyone entering the facility, visitors and workers alike, must pass through a security perimeter featuring pressure-sensitive anti-terrorist fences topped with razor wire. A black-clad platoon of wary guards checks identification before allowing anyone to pass through an imposing electronic screening device. The Bruce Power public relations officials are quick with assurances that even though the front-line guards carry no weapons, an armed response unit is at the ready twenty-four hours a day. Its members have been recruited from military combat units. The penitentiary-style security is complemented by an obvious concern about the dangers of getting an overdose of radiation. The guides carry thermoluminescent decimetres so that they can keep a sharp eye on the levels.

The world of nuclear power manifests an uneasy co-existence between a calm reassurance that all is well and an obsessive concern not to let dangerous people in or deadly radiation out. Yet aside from increased anxiety about the threat of terrorism, not much has changed since 1979, when the protestors occupying the Darlington site on the Lake Ontario shore argued that nuclear power generated too much radiation that would remain deadly for far too long.

Some twenty-five years later a new windmill—or wind turbine, to the alternative energy set—stands on the Lake Ontario shore in the heart of downtown Toronto. Imported, appropriately enough, from Holland, it is thirty stories high, which means not only that it can catch the wind off the lake but also that drivers rushing by on the nearby Gardiner Expressway cannot miss it. The thing looks elegant, its huge blades spinning lazily in the breeze. Even though, with its 24-metre blades, it assumes monstrous proportions from up close, the contraption's size pales in comparison to the bulk of an Ontario nuclear plant.

If the three blades—each the length of a single wing on a 747 airliner—make any noise, the nearby traffic's steady roar drowns it out. The windmill site resembles a suburban backyard, with a wooden deck, barbecues, and picnic tables surrounding what is proudly proclaimed to be North America's first urban wind turbine. The only sign of security is the lock on the oval-shaped door at the tower's base. The greatest danger on the site appears to be the possibility of getting dizzy from watching the blades spin. Weatherproof information panels inform the curious visitor about the basics of climate change, air pollution, energy efficiency, wind power, and co-operatives. The historically minded can stroll over to the nearest building, Scadding Cabin, built in 1794 and moved to its present location by the York Pioneer and Historical Society in 1879. An old-fashioned windmill once sat on the grounds nearby, but it was demolished to make way for the expressway at about the same time that Ontario Hydro was dreaming of building nuclear reactors.

Toronto's new waterfront windmill was started up by a non-profit group called TREC, Toronto Renewable Energy Co-operative. Lawyer Brian Iler, one of the group's organizers, was at the Darlington construction site during the times of protest and had acted as counsel for arrested demonstrators. Many of the protestors back then had sported solar-power T-shirts displaying sunny red happy faces: "Darlington? No Thanks." Iler recalled that the original idea of building an urban windmill that would offer a demonstration of a different direction had originated with Dan Leckie. The former city counsellor had not lived to see the blades start spinning, but a sign on the site indicates that the generator is dedicated to his memory. Dan Leckie had said, "Why don't we put a wind turbine in downtown Toronto? It would be highly visible as a symbol of alternative energy. It would be an example of producing locally and consuming locally."

This sort of bioregional thinking had instant appeal, and the organizers easily found investors ready to help put the thing up. They also got an enthusiastic reception from their local utility, Toronto

Hydro, which agreed to enter into a fifty-fifty partnership with TREC to build the windmill and help market the power. Then they had to leap the regulatory hurdles required to get the machine connected to the grid—no Ontario government had ever actively sought to make things easy for renewable energy boosters. The TREC windmill began spinning in early 2003, and although the market that opened briefly in 2002 might in theory have provided an opening, renewable energy was only a sideline for the privatizers. Still, many greens were fed up with the old Ontario Hydro.

"I hated the idea of dismantling a publicly owned utility," Iler said. But from his point of view Ontario Hydro had botched things up so badly, no one was prepared to defend the Crown corporation anywhere. When Ernie Eves clamped on the Remembrance Day price freeze and effectively abandoned the privatization experiment, the question was, where to turn? How to meet Ontario's energy needs without burning the planet or the provincial credit card?

The old left had a name for this type of problem, something a clever Italian journalist/politician dreamed up. Sitting in one of Benito Mussolini's jails and pondering the Great Depression of the 1930s, general misery, and what appeared to be looming war, Antonio Gramsci was trying to figure out what had gone so utterly wrong. He came up with the theory of an "organic crisis"—a time when "the old is dying and the new cannot be born; in this interregnum a great variety of morbid symptoms appear." The organic metaphor seemed particularly appropriate for the Ontario case. Among the morbid symptoms exhibited by the province's power system were air unfit to breathe, rivers poisoned with uranium mine-tailings, lakes soured by acid rain, an overheating planet, and disappearing species.

The Remembrance Day reversal showed the depth of the crisis—though it was not a total reversal, for the market would remain open, with the price cap hiding the effects of the policy from consumers. This condition prevented a ratepayer revolt but did nothing to solve the system's long-term problems. The reversal, coupled with the collapse of Enron, sent the investment climate for electricity generators into the economic equivalent of an ice age as private-sector companies packed up and went home. The government couldn't sell a generating station unless it was willing to accept rock-bottom prices.

The problems plaguing the old model had not gone away. Most of the power produced in Ontario had been effectively squandered. Only a fraction of the heat energy produced in the province's massive coal and nuclear plants was transformed into electricity and made its way into homes and factories, where most of it was being used in inefficient equipment. Little was being done to try to reduce this waste. Ontario Power Generation had poured over a billion dollars into the Pickering station and was not even close to getting the first reactor back on-line, much less all four. And even if it did get them all restarted, there was still no answer to the old question of what to do with the radioactive waste. The utility's coal plants were under attack for inflicting asthma attacks on today's kids while changing tomorrow's climate.

The anti-privatization forces faced a novel challenge: success. Their supporters thought they'd already won. Lacking even modest victories during the Harris years, many opponents now cheered the retail price freeze, the decision to keep Hydro One in public hands, and the government's inability to sell off any more plants. "Last fall, Eves faced a political crisis and an electricity crisis," Paul Kahnert would repeat at every stop on his second wintertime provincial tour in as many years. "He solved his political problem with the rate cap, but has made the electricity crisis worse."

The public now saw no "electricity crisis," and Kahnert's Ontario Electricity Coalition faced an uphill battle to get across the message that Eves was picking voters' pockets to keep the cash flowing to private electricity producers. Most people assumed that the fixed rate meant that the market was closed. But private power producers were still getting the market price. In the first year after the spot market was opened, the average price of power was 6.2 cents per kilowatt-hour, though the rate paid by households and small businesses was retroactively capped at 4.3 cents—the same price that the NDP had frozen it at nearly ten years previously. The difference between what consumers paid and private producers received between market-opening and November 1, 2003, was $1.74 billion. The terms of the market mitigation agreement required OPG to refund most of its revenues above 3.8 cents per kilowatt-hour, so that the net cost was a mere $800 million—probably more than the government of Ontario had spent on deliberate energy conservation measures in the history of the province. The $800 million was covered by the Ontario Electricity Financial Corporation.

These hundreds of millions in increased debt would have been much higher but for the subsidy coming from the publicly owned Ontario Power Generation. OPG produced 70 per cent of the province's power, but under the terms of the 1998 electricity privatization legislation it could not receive the full market price until it sold off half of its remaining assets. The company (and hence the province) went further into debt as it sold below its cost of production, but the Eves government tried to put this problem down to OPG mismanagement, ignoring the role of government policy. For years the Tories had been claiming that they had to privatize Hydro to pay off the debt. Now they were pulling out OPG's credit card to help balance their books.

The Remembrance Day reversal had the remarkable effect of ensuring that everyone involved in the system received the wrong economic signal: consumers were being told to increase consumption (because the cost of power was being subsidized), and would-be private investors were being driven out of the province by what they recognized as an unsustainable pricing policy. Meanwhile, the government rediscovered the public sector, announcing that it would invest in new capacity at Niagara and go into partnership with a private company for a gas plant in Toronto. But the power would still have to be sold at a loss. The result would probably be blackouts as demand rose and supply stagnated. The Independent Electricity Market Operator had started underlining this possibility with regular reports on the dire state of the province's power system and the potential for shortages. Few Ontarians, however, were reading IMO reports or had even heard of the Ontario Electricity Financial Corporation. With the lights on and prices low, they were largely satisfied. The political half of Eves's electricity headache went away.

Sort of: for while electricity prices faded from the headlines, its handling of the issue haunted the Eves government. Not only had the befuddled Hydro policy cost the Conservatives their reputation as good economic managers and alienated their backers on Bay Street, but the legal defeat of the Hydro One sale had also cost the government a one-time influx of $5 billion into its coffers—a payment that would have enabled them to balance the books while offering tax cuts and various goodies to voters. Ultimately, the sale, if it had gone through, would have cost the voters much more in foregone revenue and higher electricity prices, but the game of politics is played in the short term. Without that bag of treats and with the added public relations disaster of releasing a budget in a car-parts plant owned by one of the party's major donors, Tory re-election prospects were looking decidedly grim. A widely anticipated spring 2003 election call was cancelled, and the government held its breath, hoping that the lights would stay on during the hot summer days.

They didn't.

ⓦ

One graffiti artist was inspired by the Beatles' song "Blackbird." "All your life, you have only waited for this moment to be free."

The words were set against the darkened skyline of downtown Toronto, with stars forming a soaring dove on the side of an otherwise innocuous shop on Dundas Street West. There was a headline, too: "Blackout 2003."

The image captured something of the magic many people felt on the evening of August 14, 2003, when the stars came out in cities across most of the Northeastern United States and Southern Ontario. Bars stayed open late, urging people to drink the beer while it was still cold. Couples strolled arm in arm through the streets, gazing up at a night sky they could almost always never see because the stars were usually obscured by the glare from so much excess light. People stopped to chat on the candlelit porches of neighbours they had never met. Everyone suddenly had something in common—no one had power. The feared crime wave never materialized, and no lives were lost as a direct result of the blackout. Everyone had something to talk to strangers about, and there was a mass outbreak of co-operative feeling as ordinary people stepped up to direct traffic or check on elderly neighbours. The immediate reaction was to enjoy the break as much as possible and congratulate each other on behaving so well.

When the lights didn't come back on immediately, a far less romantic reality set in. Rotting food in useless fridges, the very real possibility of hospitals running out of backup power, or the prospect of cities quickly running out of water took some of the glow off the moment—particularly in Ontario, where it took far longer for the system managers to get the power back on.

The Great Blackout of 2003 underscored the fundamental importance of electricity to twenty-first-century North America. Some fifty million people were left without power, and everyday life came to a screeching halt. In the space of nine seconds, the massive cascade of electrons flowing into Ontario from the United States to run the province's air conditioning units reversed direction. It was the electrical equivalent of the Niagara River running backwards up the Falls, blowing circuits and tripping emergency shutdown systems at power plants that can't handle sudden reversals and the fluctuations in current that accompany them. In all, 263 power plants with 531 generating units capable of producing over 60,000 megawatts of electricity-generating capacity lay useless. There was no way of transporting the electricity to customers, and many generators were disabled by the sudden current swings.

Premier Eves, desperate for a chance to look like a Rudy Giuliani-style leader, popped up regularly at press conferences to reassure the public that everything possible was being done to get Ontario's electricity generators humming again. He declared a state of emergency, asked all industrial users to cut consumption by half, and despite the withering heat pleaded with citizens not to turn on their air conditioners.

While the United States largely returned to normal within days, most of Ontario's nuclear reactors remained stubbornly shut down. Of the eleven nuclear reactors that were operating at the time of the blackout, four were able to throttle back on electricity production—essentially go into an idle mode—and were soon able to feed power back onto the grid. But in the remaining seven, emergency shutdown systems dropped control rods into the reactor, absorbing the overexcited neutrons that keep the nuclear chain reaction going and work to kill the nuclear reaction before it can spiral out of control: or so goes the theory. In three of the reactors at the Pickering station, it wasn't clear that the control rods had worked. As a result the last line of defence was triggered, and the reactors were "poisoned" to make sure that they really were dead. Gadolinium nitrate gas was released into the reactor chamber, stopping the reaction but also necessitating a lengthy decontamination process that delayed the restart of the reactors by an extra week.

The blackout highlighted the fragility of a system built around large, centralized generating systems and a network of high-voltage transmission lines. The early debate on what caused the blackout focused on the role of First Energy, a particularly unscrupulous private power company in Ohio with a history of cutting corners to make an extra buck. First Energy neglected to cut the trees under its power lines, so that as the lines sagged under the increased demand on a hot summer day, they came into contact with the trees and shorted out. The company also made a number of control-room mistakes and failed to warn others of the impending collapse, not least because its computerized alarm system had shut down and its workers hadn't noticed. It didn't help that one of the state regulators turned off his alarm system to deal with a problem; he forgot to turn it back on and went to lunch.

Nevertheless, such mistakes were not particularly unusual. Human error is, after all, human. It should not leave millions of people in the dark and, for most of the last century, it hasn't. Placing the blame on First Energy appealed to the official U.S.-Canada Power System Outage Task Force because it created the idea of a single villain, reassuring the public that it was all just a one-time glitch. But First Energy was only the straw, and the camel in this case was a vastly overloaded electricity grid. Blaming First Energy for betraying the public trust ignores another fact. Loosely regulated private power companies are not beholden to the public. They owe their fealty to their shareholders. For them, gambling with the grid by cutting corners is a rational act.

Moreover, deregulation itself was destabilizing the grid. In the opening up of the long-distance transmission system, buyers and sellers of electricity become physically much further apart, and no one wants to pay for the upgrades necessary to ensure safe operation of the long-distance system. The complex electrical grid requires extensive oversight to ensure that supply matches demand on a moment-by-moment basis. With no rules governing behaviour, only voluntary guidelines, this is a rather tricky business. As David Cook of the North

American Electricity Reliability Council (NERC) told the task force, "In the absence of appropriate U.S. legislation and complementary Canadian actions, NERC continues to suffer from a lack of legally sanctioned authority to enforce compliance with its standards."

The grid requires regular maintenance, but in the deregulated markets maintenance was slashed as competing companies cut costs. Donald Wightman of the Utility Workers Union of America told the task force that his union had recently "observed a trend of declining maintenance coinciding with dramatic cutbacks in electric utility staffing. . . . No utility can provide safe and reliable service unless it has an adequate complement of utility workers." Indeed, many companies had apparently heeded Enron's CEO Jeffrey Skilling's advice to "Get rid of people. They gum up the works." North American companies got rid of more than 160,000 workers to increase profit margins—more than 16,000 in Ontario alone—leaving the general population more vulnerable than ever to interruptions.

When Ottawa selected its appointees to the task force investigating the blackout, the only non-governmental representatives were long-time Tory privatization booster David McFadden and Bruce Power supremo Duncan Hawthorne. It came as little surprise when the panel ignored evidence on the effects of deregulation presented by the North American Electricity Reliability Council, the Electric Power Research Institute, and the embattled First Energy.

Defending his company against accusations that it was primarily responsible for the failure of the grid, First Energy CEO Anthony Alexander argued:

> Competitive markets have pushed the grid beyond its design, while . . . virtually no major transmission projects have been undertaken in North America. The margins that were built into what once were local grid systems to accommodate changes in local load patterns and reliability have been reduced to allow power transactions over long distances to areas with inadequate supplies of local generation.

Alexander also complained about being singled out for cutting corners (or at least for not cutting trees) by pointing out that his company's tree-trimming practices were "aligned with current industry standards."

The task force did not question why the grid was so overloaded on a hot summer day. It did not consider the possibility that something (conservation measures, perhaps) could have been done to reduce demand and thus the stress on the system.

In Ontario, Premier Eves appointed Glen Wright to lead a provincial probe of the blackout, but the investigators proved to be not all that interested in digging too deeply into the policy roots of the collapse. Wright had been the Conservative Party election-tour director in 1995 when his fishing buddy Mike Harris won power. In 2001 he had been appointed chair of Hydro One. In that position, in a move calculated to distract public attention from the government's missteps, he had led the public lynching of Eleanor Clitheroe for abusing her expense account. It came to light that within months of his appointment Wright himself had spent over $5,000 of Hydro One's money to take a handful of Tory cronies, including media consultant Paul Rhodes, to an exclusive hunting club on a private island in Georgian Bay. The expenses included $300 on ammunition and $170 on pheasant pies, apparently because lunch was not included in the charge of $750 per night at Griffith Island Club. The hunters roughing it in the bush also included Don MacKinnon and Gerry Karn of the Power Workers' Union.

Not surprisingly, Wright's task force found the government blameless for the blackout. With the focus still squarely on the Americans, and Eves experiencing his first upward movement in the polls for his post-blackout, shirt-sleeve press-conference performances, the premier and his handlers enjoyed a brief spurt of optimism. Those advisers included Deb Hutton and Jamie Watt. Hutton had recently moved back to Queen's Park from a sinecure at Hydro One. Watt was an attack-ad specialist whose consulting firm received a quarter of a million dollars for untendered political polls paid for in part by Hydro One and OPG during the leadup to the failed Hydro One privatization. Hydro One, it seemed, was serving as a way station and source of cash for political technicians who had been running the Tory machine since the early days of the Harris administration. Their last, desperate kick at the electoral can would be a fall 2003 election call.

Neither the ruling Conservatives nor the Liberals, who were ahead in the polls, wanted to talk about electricity during the campaign of 2003. For the Tories, it was a topic too painful to contemplate. Instead, the government used the time-honoured political dodge, appointing a task force to study the issue. The Electricity Conservation and Supply Task Force, packed with privatization boosters, was co-chaired by Peter Budd, the principal spokesperson for the private power producers in Ontario.

For the Liberals, raising the electricity issue might remind voters of their various policy reversals and embarrassments. They had voted with the government on the Electricity Competition Act and had at one time supported the sale of Hydro One. On the same day that leader Dalton McGuinty was proclaiming his opposition to privatization in a media scrum, the party sent out a fundraising letter to electricity companies saying that only the Liberals could be trusted to follow through on electricity privatization.

The Liberals did make a number of ambitious campaign pledges vis-à-vis the electricity system. Things would be better by the end of a Liberal government's first term in 2007, they promised. They would shut down the province's five coal-fired generating stations, bring in conservation measures to reduce electricity consumption by 5 per cent, and provide enough "green" power to meet 5 per cent of the province's energy needs—all within four years. These clean-air commitments were driven personally by McGuinty, somewhat to the dismay of his advisors. He was knowledgeable about the issue and keen on doing something about climate change. The coal phase-out promise raised the environmental stakes while setting the Liberals apart from the Conservatives. The NDP under Howard Hampton matched these commitments and went further on the goal of conservation. Hampton, whose book *Public Power: The Fight for Publicly Owned Electricity* had just been published, soon issued a detailed plan on how the party would achieve its targets. Even the Tories got in on the act, promising a "green power standard." The NDP campaign, originally slated to hinge on the electricity issue, had to be changed so that "Public Power" was more broadly defined. The social democrats would campaign on a defence of the public sector and public services, particularly health care.

Reading the polls and under attack from the anti-privatization OEC, the Liberals were soon professing their opposition to the sale of any publicly owned electricity assets. McGuinty was quite clear in a mid-campaign newspaper article that appeared under his name: "After the election, we will stop the sell-off of hydro, period. Ontario consumers will buy their power at a regulated rate from their public power company." The Liberals were shrewd enough to leave wiggle room by remaining vague on who would finance, own, and operate any new generating plants needed to replace the aging nukes and the soon-to-be-phased-out coal burners. They also promised to keep the Eves government's rate cap in place until 2006. Ratepayers shouldn't have to pay for the mistakes of the previous government, they said.

In the middle of the campaign, the greens stirred the pot with a David Suzuki Foundation report calling on the new Ontario government to learn from the California experience. Following its disastrous experiment in deregulation, California managed to put an end to blackouts and price spikes, reducing overall consumption by 7.5 per cent in a single year by investing over a billion dollars in energy conservation and bringing in tough new rules on energy efficiency. Deregulation gave way to higher renewable energy targets that foresaw 20 per cent of electricity coming from the wind and the sun by 2010. California environmentalist Ralph Cavanaugh showed up to emphasize the basic green line: it's a lot cheaper to save energy than to burn it, which means that conservation isn't part of the solution, it is the solution. Ever since California's prioritized energy-efficiency programs in the 1970s, the state had avoided the need to build the equivalent of ten new power plants and saved $15.8 billion. These programs, scaled back with the abortive move towards a deregulated market, had been ramped up once again.

The Suzuki Foundation report had little impact on the election. According to the politics of electricity in Ontario, price and availability of power are central. The blackout was soon forgotten, while the Remembrance Day reversal with its frozen prices had effectively removed electricity as a major public issue. The main election issue was that ever-reliable time-for-a-change feeling, which the Liberals played successfully. After eight years the Tories appeared listless and increasingly incompetent. They had no new ideas, and the bungling of their big 1995 idea—privatizing Hydro—contributed to their malaise. On October 2 the Liberals swept to power with 46 per cent of the popular vote. The Conservatives lost over half their seats, and eleven cabinet ministers went down to defeat. Although the NDP increased its vote, the vagaries of the first-past-the-post electoral system meant that it finished with fewer seats and lost official party status.

In what was quickly becoming a ritual of Canadian politics, the new government immediately announced that the previous regime had mismanaged the books. The deficit was much higher than expected, and the Liberals said they would now not be able to keep all of their campaign promises. One of the first promises that they chose to break was their pledge to maintain the November 11 rate freeze. The cost of $800 million, they said, was simply too high. "Our plan will provide stable and predictable electricity prices for consumers, encourage conservation, create environmental benefits, and attract new sources of supply," said the new energy minister, Dwight Duncan, a month after the election. Rates would be going up within months. "We are moving forward with a responsible, sustainable approach that will put an end to unaffordable taxpayer subsidies of electricity prices."

A few weeks later Duncan got the first of a blizzard of reports that were supposed to help Ontario solve the intractable mess that was its power system. The first two had been commissioned by the Eves government, but timed to arrive after the election so that the government could brush aside questions about its electricity policy by simply responding that it had a committee looking into it.

The first probe, which looked into the cost overruns and delays at the Pickering A nuclear station, was headed by a former federal Conservative cabinet member and well-known friend of the nuclear industry. As the minister of energy, mines and resources under Brian Mulroney, Jake Epp had overseen a big increase in the budget of Atomic Energy of Canada Ltd. His Manitoba riding was home to AECL's Whiteshell Nuclear Research Establishment. In 1990 he had assured the nuclear industry that he was determined "to change the negative attitudes that surround both AECL and the nuclear industry generally." He had, he said, "put a lot of effort" into working on the task because he was convinced that nuclear energy was "the way to go." The task was daunting: "AECL and the nuclear industry don't have friends in every corner and there have been very few cabinet ministers willing to tackle this nut."

Epp had been asked to look into the problems at Pickering back in June 2003. The slim *Report of the Pickering "A" Review Panel* laid out "a catalogue of problems" involving the attempted restart of the four reactors that had been closed in 1997 for safety reasons on the advice of the U.S. consultants. It noted that the powers-that-be had started the work without a plan as to how to finish it or even the most basic idea of what order the work should be done in or who was in charge of what. Some eleven cost-estimate increases (from $780 million to $2.5 billion) and thirteen "scheduled return to service" dates later, and with only one of the four reactors working, OPG had given up offering guesses as to how much the work would cost or when it would be done. Epp helpfully offered his opinion that the cost would probably be somewhere between $3 billion and $4 billion to complete the job, if they fixed the management problems.

Duncan thanked the review panel for its work, called the effort to restart the station a "horrible mess," and promptly fired OPG chair Bill Farlinger, CEO Ron Osborne, and chief operating officer Graham Brown. Firing Farlinger was inevitable. Like the chairs of Ontario Hydro in days gone by, he served at the pleasure of the government and had offered his resignation after his premier was defeated. But the Liberals twisted the knife. The Tories had exempted OPG and Hydro One from Ian Scott's Freedom of Information Act so that the two bodies could behave like private corporations; the Liberals immediately dropped this protection. Shopping carts of material were rolled out, and before long Harris's mentor was subjected to further humiliation. The documents revealed that over five years Farlinger and Osborne had billed the public corporation for over $800,000 in expenses, which included trips to London and Tokyo. They had entered their golf cart rentals, green fees, and club polishing as expenses in Palm Springs. The social whirl took the electricity executives and their guests to private boxes at hockey games, concerts, and theatres.

After-dinner cigars at the exclusive Toronto Club cost Ontario power customers $75.

The Tory backroom boys and girls were not to be outdone. Hydro One had paid one Tory insider to offer another Tory insider a job, or that's how it seemed when Tom Long's Egon Zehnder International received $83,000 to recruit Hutton for a top job ("senior vice-president of corporate relations") at Hydro One. Hutton (known as Jabba the Hutt in the Cabinet Office) graciously accepted. Long, an author of the poor-bashing Common Sense Revolution and chair of the 1995 and 1999 Tory campaigns, once charged $650 an hour for writing Farlinger's speeches. Hutton advised Harris on the campaign trail and went on to work in the premier's office. Leslie Noble's firm, StrategyCorp, got $250,000 from Hydro One; Noble was another Common Sense Revolution author. Paul Rhodes, another member of the Harris/Eves inner circle, got $335,000 for eighteen months' work. All in all, top Tory insiders got $5.6 million in untendered Hydro One contracts. The very idea of the defenders of the common taxpayer making away with so much for so little gave new meaning to the notion of public-private partnerships. It would later emerge that Harris himself picked up $18,000 of OPG money as a reward for offering advice about David Peterson's old idea of building stronger interconnects between Ontario and Quebec.

With the departure of the disgraced free-market fundamentalists, the privatizers were in retreat. After over seven years at the helm, they could only claim one lease and the sale of four hydro dams—the value of which were coming increasingly into question. It had also been seven lost years as far as conservation and renewable energy were concerned, and the $12-billion-a-year electricity industry still faced its organic crisis, morbid symptoms and all.

The next document to land on Duncan's desk was the much heftier report of the Tory-appointed Electricity Conservation and Supply Task Force. *Tough Choices: Addressing Ontario's Power Needs* was the product of the private power cheerleaders who had produced the 1998 legislation, but by this time their enthusiasm was distinctly muted. The report opened by recognizing, "Major changes in the energy economy and in public policy have undermined the viability of the original market design." All of this had to do, the writers said, with the "demise" of financial markets following the collapse of Enron and other energy traders, the rise in natural gas prices, and the Liberal government's commitments to phase out coal and keep OPG assets in public hands. The report did manage to express a wistful support for "an electricity sector that is based on competitive principles," but acknowledged, "A market-driven system alone is unlikely to deliver the new generation and conservation Ontario needs within the time frames we need them."

The report's most controversial aspect was its attempt to reconcile the writers' free-market aspirations with the new government's commitment to a phase-out of coal. Task force co-chair Peter Budd—the free market's most public advocate after Energy Probe—was sacked by the energy minister when he mused publicly that the Liberals' pledge to shut down the coal plants was impossible to achieve by 2007. Power Workers president Don MacKinnon, who had also been assigned by the Tories to the task force, refused to sign the report because he believed it was too hard on coal and too easy on natural gas. Indeed, phasing out coal would be what people in more polite circles called a "challenge." The producer-driven task force appeared to come out in favour of creating a "conservation culture" as the cheapest and cleanest way to resolve the province's electricity woes; but the document's fine print revealed a crisis of imagination. The authors could only foresee reducing the *growth* of power demand; cutting demand itself was still inconceivable.

The idea that Ontario could realistically use less power was still apparently confined to back-to-the-land organic farmers living in straw-bale houses. The task force reinterpreted the government's pledge to reduce electricity consumption by 5 per cent by 2007 to mean 5 per cent below a ridiculously high projected growth rate of 1.7 per cent. The Independent Electricity Market Operator was only predicting 1.1 per cent, and actual growth had been only 0.7 per cent since the 1989 Demand Supply Plan was first drafted. In an alarming note, the report argued that without new generators and massive conservation, Ontario would only have the capacity to meet half its needs by 2014. The task force recommended new nuclear reactors, along with a few windmills, to meet the rising demand.

The pile of documents was soon augmented with reports commissioned by the Liberals, and those findings also sounded the alarm. Financial analysts at KPMG reported that Ontario Power Generation's financial situation was deteriorating to the point where the corporation's financial viability was in question. The cost of refurbishing Pickering A was still on the way up, with that burden augmented by the high costs of running the fossil-fuel plants and thus filling the nuclear gap. To top it off, OPG's revenues were still capped at 3.8 cents per kilowatt-hour by the market mitigation agreement. The same day that the KPMG report appeared, the government announced its decision to keep its promise of closing the coal-fired plants. They and their 7,500 megawatts of capacity would be shut down. OPG immediately released its financial results for 2003, showing a loss of $491 million, much of it attributable to closing the coal plants. Within days another report, this one by a Liberal (former Deputy Prime Minister John Manley, who retained prime ministerial ambitions), a Tory (the ubiquitous Epp, now OPG's interim chair), and a banker (a former CEO of Scotiabank), told the government that Ontario should again follow the nuclear path by pouring more money into Pickering A. It should also start pouring the cement for a fresh round of new nuclear capacity because natural gas was simply too expensive to contemplate. The new nukes could be built in whole or in part by the private sector, including foreign firms. Still, the report said, there had to be a new emphasis on conservation.

It was an odd mixture, this quicky report from an establishment panel. It reflected the crossroads that Ontario was facing: old-style thinking about a centralized grid based on nuclear power co-existed with a recognition that the state had to take action on conservation. The ideological ground had clearly shifted with the panel's recognition that privatization of existing generation capacity was simply not on. At the same time the panel recognized that Ontario finally had to bring in tough, independent regulation. "The experience elsewhere," said Manley, "is that regulation can drive better performance, whether the electricity supplier is owned by government or the private sector." He was obviously referring to jurisdictions such as California and Vermont in the days before Enron-inspired deregulation.

When the Liberals took over, the only real electricity regulation was just as it had always been back in the days of the old Ontario Hydro: it depended on the willingness of the government of the day to absorb the political fallout from higher prices. The Manley report referred to this tendency politely as "a highly political process." The open market had continued to function under the aegis of the IMO since the Tory climbdown; it had 278 participants, from casinos and paper mills to municipal utilities and OPG itself. Though most people were unaware it even existed, the IMO was still overseeing the wholesale market as suppliers submitted offers to sell electricity and wholesale buyers made bids to buy it. Private suppliers could make very tidy returns when the price rose, and rise it did, whenever the weather got really hot or really cold. Now—to keep the price increases down and the lights on—the government was being told to do some tough regulation and look to the money pit of Pickering station in the short term, while considering conservation in the medium term and new nukes down the line.

Maybe it was not true that a lethal combination of Maurice Strong and absurd costs had finally driven the stake through the heart of the Dracula that is the nuclear power industry. It did seem like another case of the environmental opposition being stymied by its own success. Having effectively demonized coal for poisoning the planet and choking the children, the greens watched in horror as their old nemesis, "the nukes," seemed poised to run away with the Comeback Player of the Year award. The day on which Ontario's New Energy Directions and its sweeping conservation and efficiency programs hit the front page, the bottom of the same page featured a full-colour ad with a clogged highway and a bar graph showing that Canadians were the world's highest per capita consumers of energy (in barrels of oil equivalent). "Test your energy IQ," urged the Canadian Nuclear Association.

The government was being sorely tested. When he delivered his report on the future of OPG, Manley proved to be more astute regarding politics than he was with power policy. Dealing with the electricity mess, he said, was the biggest issue that the McGuinty government would face in its first mandate. If anything were to go wrong, he predicted, "The political price to be paid will be by the government in office, it won't be by all the previous governments that have planned for this in the past or perhaps pursued the wrong policies."

In a much-anticipated speech to the Empire Club six months after taking office, Liberal Energy Minister Dwight Duncan warned that the province stood "dangerously close to an unforgiving precipice, which threatens to undermine Ontario's continued prosperity." The McGuinty government's approach, he assured the assembled Bay Street elite, would be one of "balance"—a word he repeated nine times in rapid succession. Duncan finally laid out the government's response to the blizzard of public reports and behind-the-scenes lobbying. The Liberals would reject the Tory model. "I want to move forward. We've looked at moving to a fully competitive market, but couldn't find one that worked—anywhere." The new approach would instead "balance public leadership with private investment."

Seemingly in keeping with McGuinty's conversion to public power during the election campaign, the Liberal government would oversee no sale of Ontario Power Generation or Hydro One assets. The publicly owned plants would feed power onto the grid on a "power-at-cost" basis, except this time the cost would be determined by a stronger and more independent Ontario Energy Board rather than by the government or the utility. The idea was to remove politics from electricity-pricing.

The future rested in creating "a climate that welcomes private investment." New generators would be privately owned and would be able to charge whatever price came out of the interplay of market forces. To soften the blow, consumers would have the option of paying a fixed rate, set annually by the Ontario Energy Board, or playing the market. In any event, because 70 per cent of the power was being supplied by the public sector at a fixed rate, prices were expected to be less volatile than they were under the completely open market. The old hydro dams that cost about a penny a kilowatt-hour could offset the seven cents per kilowatt-hour being demanded by the private sector as a prerequisite for investment.

In many ways this approach was not that different from the Tory model, which capped OPG's revenues after-the-fact through the market-power mitigation agreement and used the difference between the market price and what OPG was paid to subsidize the cost of private power. The Liberals were simply capping the price beforehand and blending the cheaper "heritage" power from Niagara Falls with the higher-priced private power. This was no renewed public power system, but a recipe for slow-motion privatization. Many publicly owned generating stations would be phased out by attrition, gradually replaced by private, for-profit stations as coal generators were closed and nuclear plants came to the end of their lives. If the Liberals did manage to shut coal down by the end of their first mandate, private operators would be supplying over half of Ontario's power within four years.

As Liberals, McGuinty and his government were still trying to be all things to all people, and their faith in the free market was far from absolute. If the private sector did not leap at the opportunity to invest, as it hadn't globally since the collapse of Enron, the government would build new plants to make sure the lights stayed on. This approach was similar to the Tory position under Ernie Eves.

What was truly different from the Harris/Eves model was a much more significant role for the government in system planning. Rather than leaving the balancing of future supply and demand up to the market, the government would set the broad targets for conservation, for energy from renewable sources like wind turbines, and decide whether or not to build new nuclear plants. Working out the details and signing the contracts with suppliers would be the responsibility of a new public agency, the Ontario Power Authority. Operating at arm's-length from the government, this new agency, the Liberals hoped, would be less vulnerable to engineering hubris than was the old Hydro, which did both the forecasting of future demand and the building and operating of the plants. To ensure that conservation did not fall off the agenda, the government proposed creating, within the Ontario Power Authority, a conservation secretariat that would be responsible for ensuring that conservation targets (initially a 5 per cent reduction in demand by 2007) were met.

The real surprise in Duncan's spring 2004 speech was what he did not announce. He made no mention of re-tubing Pickering A or the construction of new nuclear plants. With the Manley report's enthusiastic endorsement of new nuclear plants and rumours that two private-sector proponents were offering to build new reactors at existing sites, many observers had expected that Duncan would at least announce a go-ahead for Pickering A. The decisions were delayed, perhaps because the government was seeking to put off the controversy as long as possible or because it truly had not yet decided. Duncan did reiterate the government's commitment to close the coal plants by 2007, as long as there was sufficient supply to replace them.

The coal phase-out offered Ontario an important opportunity to take the first step towards a more sustainable electricity system. The huge coal closure could not be achieved by muddling through on a few changes at the margins. Achieving the phase-out would require bold new steps—a choice between the hard and soft paths not unlike the Darlington decision made by the Peterson government.

A very real possibility remained that a coal phase-out would be achieved through a repeat of the Darlington fiasco. History would repeat itself, this time as farce, as investment in new nuclear capacity trumped conservation and renewables. Still, shutting down the coal burners would open up the possibility of a dramatic turn onto the soft path. Environmentalists who wanted to nudge the government in that direction unveiled a comprehensive plan showing once again how a soft path strategy could combine two elements: an aggressive conservation strategy to cut energy consumption by 40 per cent by 2020; and a phase-out of coal and nuclear plants in favour of new renewable energy and high-efficiency gas co-generation plants. An analysis by a team of economists from Simon Fraser University found that this conservation-driven strategy would save $14 billion in capital costs relative to building new nuclear plants. As an added bonus, 96 per cent of the $18 billion in capital costs for the investments in energy efficiency would be recouped through energy cost-savings.

These economic and environmental benefits would be spread relatively thinly but equitably right across the province. In contrast, profits from building and operating new plants—whether nuclear or fossil—would be concentrated among a few corporations. The path that the McGuinty government chose would shape Ontario's electricity future.

It had been a hundred years since James Whitney's Conservatives, another government promising to sweep corruption from the halls of Queen's Park, had taken over in Ontario. Electricity was at the top of its agenda, too. A century earlier the party had consolidated a base in the province's fast-growing cities, where the promise of electric power was on everyone's mind. The Tories rode to power on the back of a swaggering public power movement, and when Whitney arrived in office he knew that the leader of that movement, who happened to be the most influential member of his caucus, was expecting big things. That man was named Adam Beck and the big thing would turn out to be Ontario Hydro.

As soon as he took office, Premier Whitney repeated a phrase that he had often used during the campaign. The electricity from the storied Falls at Niagara should be "free as air." It was easy for the average voter to grasp those three little words. But like so many other catchy lines, such as Common Sense Revolution, the words masked a number of crucial assumptions. A century later a growing number of people had come to understand that if the air—and the rest of the biosphere—are treated as free goods, that means trouble: trouble in the form of air that is dangerous to breathe and radioactive garbage so poisonous that no living thing can safely go anywhere near it for tens of thousands of years.

Very few were thinking along those lines back in 1905 when Beck got down to the business of harnessing Mother Nature at Niagara. His model of a giant utility with equally large water-powered generators worked very well. It fuelled Ontario's industrial revolution for the first half of the twentieth century, helping to establish the province as Canada's richest, its manufacturing heartland. Just about everyone got electricity at home in the process. Then along came coal and, increasingly, nuclear power. During the forty-three years of Tory rule, Ontario Hydro became an even more effective tool of the state. It fit in very nicely with the principle of running an economy based on mass production and mass consumption ("Live Better Electrically"). Not only that, but spending public money on new megaprojects helped to iron out the ups and downs of the business cycle while allowing the government of the day to keep its friends in the gravel, cement, construction, and manufacturing industries happy.

The demise of Keynesianism in the 1970s coincided with the rise of environmental consciousness and left Ontario with a choice. It could listen to the green message and take the road less travelled, the soft energy path; it could keep on building more big power plants; or it could do nothing. For some fifteen years it had chosen the last option. It could do so because power demand flatlined as the economy's energy productivity improved. Meanwhile the world witnessed the rise of the idea that deregulation, privatization, and the supposed freedom of the market were just the ticket for taking care of matters that had formerly been the preserve of the state, including the provision of power. This notion took root in fascist Chile and Thatcher's Britain before making its way to North America, where it was picked up by Enron, Mike Harris, and the dominant opinion-makers of the day. Dressing the approach up as a Big Idea masked what it really was: a mechanism to satisfy personal and corporate greed.

It was soon discovered that market-driven power systems not only lined the pockets of their many backers, but also, particularly when unregulated, were every bit as likely as public behemoths such as Ontario Hydro to ignore the environmental costs of their operations. For them, all of the planet was as "free as air." What's more, a market-oriented electricity system is even more likely than a state monopoly sector to ignore long-term planning while at the same time discounting the need for public benefits.

The market can be a wonderful tool for producing material goods and mechanisms, including wind turbines and solar panels. But left to its own devices, without adequate political control, laissez-faire capitalism has an inevitable tendency to become "disembedded" from both human society and the natural environment, particularly when given free reign. The guiding spirit of such enterprise, in the words of economic and social historian Karl Polanyi, is a "veritable faith in man's secular salvation through the self-regulating market."

Polanyi's influential book on the political and economic origins of twentieth-century society was first published in 1944, the year that the Bretton Woods conference marked the formal beginning of a Keynesian era intended to guide and regulate the market, saving capitalism from itself. Ontario Hydro had already launched its Department of Sales Promotion. By the end of the century Keynesian ideas had in good measure been eclipsed by a resurgent—and particularly virulent—strain of laissez-faire. Some called the phenomenon globalization. Its advocates insisted that there was no alternative. But Polanyi had already showed how the market inevitably erodes the basis of its own operation, which includes both society and the natural world. He came up with the idea of a "double movement" that operates in constant tension: one impulse proceeds under the banner of laissez-faire and free trade; the other side is a historic countertendency—"the principle of social protection aiming at the conservation of man and nature . . . relying on the varying support of those most immediately affected by the deleterious action of the market."

At the end of January 2004, just as the Liberals struggled with Ontario's electricity conundrum, a group of people gathered at Friends House, downtown Toronto's Quaker sanctuary. The crowd of some seventy-five people included the Sierra Club's Dan McDermott, who had parachuted into the Darlington site twenty-five years before. Jack Gibbons of the Ontario Clean Air Alliance, and the person as responsible as anyone for the decision to close the coal plants, was also there. David Martin, formerly with the Nuclear Awareness Project and later also with the Sierra Club, had pulled the event together in order to put forward a green alternative vision for electricity in Ontario.

But it wasn't only an old guard of greying greens. Deb Doncaster, who had conducted the tortuous negotiations that got the Toronto Renewable Energy Co-operative's windmill running on Toronto's waterfront, attended on behalf of the Ontario Sustainable Energy Association. OSEA members were doing their best—in the face of indifferent public policy—to build the windmills, solar hot-water heaters, and micro-hydro plants that had been talked about for so long. Representatives of co-operatives that were putting up little windmills or solar installations had come to the meeting to talk about how community-based power fit into Ontario's energy future. The Toronto Hydro workers Charlene Mueller and Paul Kahnert were also there. Few in the crowd knew it, but they had been the driving forces behind the creation of the Ontario Electricity Coalition, as responsible as anyone for stopping the privatization express. Along with a handful of other labour activists they wanted to remind the greens about jobs, and about the perils of letting the price system take care of conservation in a province where the poor spend far too much of their meagre incomes on electricity.

The crowd had assembled to hear Ralph Torrie hold court about the watershed moment in Ontario—about what really was an organic crisis, morbid symptoms and all. Torrie delivered a sardonic speech on how Ontario could avoid the nukes and the coal while supplying everyone with enough power. He insisted, as he and Amory Lovins and the rest had been insisting since the 1970s, that nobody wants fuel or electricity. What everyone wants is to be able to eat a hot meal in a well-lit house that is neither too hot nor too cold.

Working his way through six-dozen slides and various graphs and tables in a cross between a revival meeting, a stand-up comedy routine, and a physics lecture, Torrie laid out the historical basis for the crunch in the electricity system. Insisting that the soft energy path was just as promising as ever, he uncovered the vast stores of energy service available through better building codes, building retrofits, solar water heaters, high-efficiency air conditioners, appliances and motors, smarter lighting, and the co-generation of heat and electricity. Then there was the possibility of harnessing the power of wind, sun, and moving water, as well as of using natural gas more efficiently, to produce both electricity and useful heat. It all added up to the old saw about opportunity bred by crisis. Ontario could phase out both coal and nuclear plants by 2020 simply by taking advantage of existing technology. The key, again, is not simply building new supply but using energy more wisely. Torrie's data showed how this was already happening. Between 1993 and 2002, a period when Ontario was not really trying to improve energy efficiency, it was doing more with fewer kilowatts. In fact, electricity productivity had emerged as the largest source of new supply—three times as large as the reduction in power production from the province's failing nuclear plants. Torrie's graphs showed that the productivity of Ontario's Candus just kept on declining as they aged.

It did not matter how many more billions were poured into Pickering, Bruce, and, soon, Darlington. They were like that old car: another cheque for the brakes and yet another to patch the rusting body did not banish that nagging feeling that the beater would some-day come to a halt, having outlived its usefulness. For the nuclear plants, that day was coming fast, but still the nuclear industry was telling anyone who would listen that it had turned over a new leaf. If only the private sector could be brought in as a partner, fresh supplies of nuclear energy would take care of all the province's power supply needs. The nuclear lobby trumpeted its new Candu reactors in China, where the product had come in "on time and on budget." China, with its sweatshops, graveyard coal mines, rubber-stamp legislature, and police state had no worries about uppity critics and bothersome safety regulations.

The romance with nuclear power had led to the decline of Ontario Hydro's electric empire. Torrie had charted its fall, but avoided the temptation to say, "Told you so." Instead, he asked, imagine what we could do if we finally did try to make the shift to a soft energy path: not just as a means of saving the planet, but also as a road leading to a better way to live? He closed by reminding his audience of the four laws of ecology: everything goes somewhere; everything is connected to everything else; there is no free lunch; nature bats last.

"Human beings are unique, in that they can make self-conscious choices," Torrie said. "We've made some pretty bad ones when it comes to how we produce and use energy. But what could we do if we apply our talents towards building a sustainable energy future?"

Some twenty-five years before the green energy crowd gathered at the Quaker meeting house, a similar vision had prompted Joe Vise and sixty-five others who had chosen the path of resistance by going over the barbed wire at Darlington. A quarter-century on, all the tensions of the double movement are still very much in evidence. The Nuclear Association's atomic visionaries continue to offer up assurances that we can have it all, and more—clean air, a healthy planet, *and* a world of endless expansion—a limitless vista of big box shopping. The countermovement insists on holding up a mirror to the past, glimpsing there the death of the sensible. The energy vision-aries insist that there is, in fact, an alternative future. This time the soft path advocates are busily building alternatives, one windmill at a time, and bringing both local utilities like Toronto Hydro and mas-sive outfits like Hydro Québec along. Their message has always been that we can indeed use less energy and employ what we do use more creatively. Their understanding is that we need to reinvent power systems so that we are working with—rather than against—nature.

A Chronology of Ontario's Electricity System 1884–2003

1884   Peterborough becomes the first city in Canada to have electric streetlights.

1906   The Hydro-Electric Power Commission of Ontario is established as a publicly owned utility, with Adam Beck as chairperson. HEPCO (known as Ontario Hydro) began as a transmission and distribution utility, but gradually built or bought a large number of hydro dams. As of 2004, Ontario Power Generation (a successor company) still owned and operated 37 large dams and 29 smaller ones capable of generating 6,928 megawatts.

1921   Completion of Sir Adam Beck No. 1 hydro plant at Niagara Falls. At 1,123 MW, it is the world's largest hydro plant.

1951   Richard Hearn Plant in Toronto is Canada's first 100-MW coal-fired generating station. The Hearn station was converted to natural gas in 1971 and closed in 1983.

1953   J. Clark Keith coal-fired generating station is built in Windsor.

1958   The 1,328-MW Sir Adam Beck 2 hydro plant is built on the Niagara Gorge adjoining Beck 1.

1957   Development begins on NPD-2, the first Candu nuclear reactor.

1958   Completion of the 912-MW R.H. Saunders hydro plant in Cornwall.

1960   Construction begins on the first medium-scale (200-MW) Candu reactor at Douglas Point, near Kincardine on the Bruce Peninsula.

1962   The Lakeview coal-fired station in Mississauga is Canada's first 300-MW thermal plant (eventually expanded to 1,140 MW).
The NPD-2 Candu reactor attains full power.

1963   Completion of the 310-MW Thunder Bay coal-fired generating station.

1965   Construction begins on the first large-scale Candu reactors (Pickering A, unit 1, and Pickering A, unit 2, each capable of generating 515 MW) at Pickering, just east of Toronto.

1967   Douglas Point Candu reactor generates its first electricity.
Construction starts on 515-MW Pickering A, unit 3.

1968   Construction starts on 515-MW Pickering A, unit 4.

1969   Completion of the 1,975-MW Lambton coal-fired station.

1970   Construction starts on Bruce-2 Candu reactor near Kincardine. The Bruce A station will eventually house four reactors capable of generating 825 MW each. The Bruce B station will house four reactors capable of generating 840 MW each.

1971   Pickering-1 and Pickering-2 generate their first electricity.
Construction starts on Bruce-1.

1972   Completion of the 3,920-MW Nanticoke coal plant.
Pickering-3 generates its first electricity.
Construction starts at Bruce-3 and Bruce-4.

1973   Pickering-4 generates its first electricity.
The Hydro-Electric Power Commission of Ontario is restructured and formally renamed Ontario Hydro.

1974   Construction starts at the 516-MW Pickering B-1 nuclear reactor.

1975   Construction starts at the 516-MW Pickering B-2.

1976   Construction starts at Pickering B-3 and Pickering B-4 (516 MW each).
Bruce A-2 generates its first electricity.

1977   Bruce A-1 and A-3 generate their first electricity.

1978   Construction starts at Bruce B-1 and Bruce B-2.
Bruce A-4 generates its first electricity.

1979 Construction starts at Bruce B-3 and Bruce B-4.
Economic and safety concerns arising from the Three Mile Island accident lead to the cancellation of new nuclear projects in the United States.

1982 Construction starts at the Darlington Candu reactor, unit 1, and Darlington, unit 2 (881 mw each).
Pickering B-1 generates its first electricity.

1983 Pickering B-2 generates its first electricity.
Coolant leak at Pickering plant forces a shutdown of reactor.
Richard Hearn fossil-fired generating station on Toronto's waterfront is closed.

1984 Douglas Point shuts down.
Bruce B-1 and B-2 generate their first electricity.
Pickering B-3 generates its first electricity.
Construction starts at 881-mw Darlington-3.
Windsor's J. Clark Keith coal-fired generating station is closed.

1985 Construction starts at 881-mw Darlington-4.
The 215-mw Atikokan coal-fired generating station begins operation.

1986 Pickering B-4 generates its first electricity.
Bruce B-3 generates its first electricity.
The Chernobyl nuclear generating station in Ukraine explodes and burns. Radioactive material is dispersed over Northern Europe.

1987 Bruce B-4 generates its first electricity.

1989 Ontario Hydro produces its 25-year Demand Supply Plan calling for 10 new Candu nuclear reactors and 32 new fossil-fired generators.

1990 Darlington-2 generates its first electricity.
Darlington-1 generates its first electricity.

1992 Darlington-3 generates its first electricity.

1993 Darlington-4 generates its first electricity.
Ontario Hydro withdraws the Demand Supply Plan from the Environmental Assessment process in the face of an excess of power-generating capacity.

1995 Bruce A-2 is shut down due to damage from a lead blanket left inside the reactor during routine maintenance.

1997 All four units at Pickering A and the remaining three units at Bruce A are closed following a critical report from consultants hired by Ontario Hydro; the report cites safety concerns and management problems.

1998 The Ontario legislature passes the Energy Competition Act, opening the way for electricity privatization and deregulation.
Ontario Hydro is broken into five successor companies, including Ontario Power Generation, which owns the generating stations, and Hydro One, which owns the transmission system.

2001 The eight nuclear reactors at the Bruce station are leased to a private consortium controlled by British Energy.
The Ontario Electricity Coalition launches a campaign against privatization and deregulation.
Premier Mike Harris announces that Hydro One will be privatized and the electricity market opened to competition in May 2002.

2002 Ontario Power Generation announces sale to Brascan of four hydro dams capable of generating 490 MW.
Superior Court Justice Arthur Gans rules that the Ontario government has no legal authority to sell Hydro One.
The electricity market is opened to competition, with prices to be set by the market.
Some 42 Ontario city and town councils pass resolutions opposing the opening of the market.
Higher electricity bills spark political anger. Premier Ernie Eves announces electricity price cap.

2003 Premier Eves announces that Hydro One will not be sold.
In the summer the largest blackout in North American history hits the Northeastern United States and Southern Ontario.

Notes

1 Introduction: The Rise and Fall of an Electric Empire

p.2, It "was known around the world": Interview with Allan Kupcis, March 17, 2004.

p.2, ". . . fallen on his sword": *The Toronto Star*, Sept. 27, 2003.

p.2, "The passing of Ontario Hydro": Interview with Allan Kupcis by Deborah Irvine, Feb. 28, 2002.

p.3, On the tour he noticed: Interview with Floyd Laughren, June 12, 2001.

p.3, Hard-wired to its corporate DNA was a belief: Richard Manning, *Against the Grain: How Agriculture Has Hijacked Civilization* (New York: North Point Press, 2004).

p.3, That is what another physicist, Joe Vise: Interview with Joe Vise, Sept. 15, 2003.

p.7, Almost thirty years later Torrie was still livid: Interview with Ralph Torrie, Sept. 17, 2003.

p.7, Instead of laughter or tears, Torrie chose the therapy: Letter to the editor, *The Globe and Mail*, March 24, 2004.

2 Power and the Tories: The Making of a Crisis

p.9, The dour British magazine *The Economist*: *The Economist*, April 14, 1979.

p.10, Just down the shore from Darlington: Penny Sanger, *Blind Faith* (Toronto: McGraw-Hill/Ryerson, 1981).

p.12, The story goes that Admiral Hyman Rickover: Interview with David Runnalls, Toronto, Sept. 18, 2003.

p.12, The journals were classified: Interview with David Runnalls, Sept. 18, 2003.

p.12, The 20-megawatt NPD partnership: Atomic Energy of Canada Limited, *Canada Enters the Nuclear Age: A Technical History of Atomic Energy of Canada Limited* (Montreal and Kingston: McGill-Queen's University Press for Atomic Energy of Canada Limited, 1997), p.395.

p.13, Hydro was, according to one background paper: <http://www.cns-snc.ca/history/npd/historical_backgrounder.html>. The paper was done for the historical plaque unveiled at Rolphton in 2002.

p.15, The two biggest areas of engineering and construction: Kenneth J. Rea, *The Prosperous Years: The Economic History of Ontario 1939–1975* (Toronto: University of Toronto Press, 1985), pp.84, 53.

p.15, No hungry man who is also sober: J.K. Galbraith, *The New Industrial State*, 2nd ed. (Harmondsworth, England: Penguin, 1974), p.24.

p.16, In 1958 AECL got a boost: Robert Bothwell, *Nucleus: The History of Atomic Energy of Canada Limited* (Toronto: University of Toronto Press, 1988), p. xvi.

p.16, "We've suffered enormously in this country . . .": *Saturday Night*, March 1977.

p.17, Now, with the Rolphton demonstrator up and running: quoted in Paul McKay, *Electric Empire: The Inside Story of Ontario Hydro* (Toronto: Between the Lines, 1983), p.41.

p.17, One day Davis and his mentor stood: quoted in Neil Freeman, *The Politics of Power: Ontario Hydro and Its Government 1906–1995* (Toronto: University of Toronto Press, 1996), pp.115–16.

p.19, In the following month Premier Davis announced: Freeman, *Politics of Power*, p.155.

p.19, Despite its own role in actively encouraging power demand: Ontario Hydro, *Long Range Planning of the Electric Power System*, report no. 556-Sp, 1974, quoted in McKay, *Electric Empire*, p.57.

p.19, The answer from McKeough was to the point: Legislative debates, June 18, 1973, quoted in Lawrence Solomon, *Power at What Cost: Why Ontario Hydro Is Out of Control and What Needs to Be Done about It* (Toronto: Energy Probe Research Foundation, 1984), p.22.

p.20, While the Best and the Brightest working at the Chalk River lab looked inward: Bothwell, *Nucleus*.

p.21, "We ran the gauntlet of construction workers": Interview with Paul Gervan, Sept. 17, 2003.

p.21, Is this wise planning? Is it planning at all: Ralph Torrie and the Ontario Coalition for Nuclear Responsibility, "Half Life: Nuclear Power and Future Society," mimeo, Infoearth, Ottawa, 1977, revised 1980, p.220.

p.23, Torrie's introduction quoted: Dale G. Bridenbach, letter of resignation, Feb. 2, 1976, quoted in Torrie, "Half Life," p.1.

p.24, If the system is remotely and impersonally managed: Science Council of Canada, *Canada as a Conserver Society: Resource Uncertainties and the Need for New Technologies*, report no.27, Ottawa, September 1977, p.32.

p.24, P.E.I. premier Alex Campbell reacted angrily: quoted in Sandra Gwyn, "Pioneers of Elegant Frugality," *Saturday Night*, March 1977.

p.25, Some three years before Three Mile Island he was pointing out: Amory Lovins, "Energy Strategy: The Road Not Taken?" *Foreign Affairs*, October 1976.

p.25, Even before these warnings appeared: Interview with Arthur Cordell, Nov. 26, 2003.

p.26, In an electrical world, your lifeline comes: Lovins, "Energy Strategy," pp.92, 86.

p.26, As the Science Council lamented: Science Council, *Canada as a Conserver Society*, p. 45.

p.27, Progressive Conservative stalwart Darcy McKeough: cited in Donald MacDonald, *The Happy Warrior* (Toronto: Fitzhenry and Whiteside, 1988), p.308.

p.28, The Commission recommended: Freeman, *Politics of Power*, pp.162–65; McKay, *Electric Empire*, pp.170–72.

p.29, The cancellation even included plants: Walt Patterson, *Transforming Electricity: The Coming Generation of Change* (London: Royal Institute of International Affairs, 1999), p.73.

p.30, This new Board of Industrial Leadership and Development would: BILD, *Electricity: Realizing the Potential*, January 1991, quoted in McKay, *Electric Empire*, p.184.

p.31, "Davis had no fresh vision to offer": Rosemary Speirs, *Out of the Blue: The Fall of the Tory Dynasty in Ontario* (Toronto: Macmillan, 1986), p.9.

3 The Electric City and Public Power: Promise, Peril, and Peterson

p.32, "Sometimes wake-up calls like this help everybody: *The Globe and Mail*, Aug. 16, 2003.

p.33, As one writer noted, "It was not by coincidence . . .": Thomas Parke Hughes, "Thomas Alva Edison and the Rise of Electricity," in *Technology in America: A History of Individuals and Ideas*, ed. Carroll Pursell (Cambridge, Mass.: The MIT Press, 1981), p.123.

p.34, In 1890 the city fathers told the Edison General Electric Company: E. Jones and B. Dyer, *Peterborough: The Electric City* (Peterborough: Windsor Publications, 1987), p.40.

p.35, A Toronto *Globe* Saturday special on "Power, Light and Heat": quoted in H.V. Nelles, *The Politics of Development: Forests, Mines and Hydro-Electric Power in Ontario, 1849–1941* (Toronto: Macmillan, 1974), pp.218–19. Much of what follows owes its understanding to Nelles's analysis of the development of Ontario's resources, a work of political economy unrivalled for the depth of its research and understanding of these issues.

p.37, Adam Beck was, in the words of one Hydro history: Merrill Denison, *The People's Power: The History of Ontario Hydro* (Toronto: McClelland and Stewart, 1960), p.33.

p.38, According to H.V. Nelles's definitive history of Hydro's origins: Nelles, *Politics of Development*, p.232.

p.38, Pork packer Joseph Flavelle: W.R. Plewman, *Adam Beck and the Ontario Hydro* (Toronto: The Ryerson Press, 1947), p.73.

p.38, Their investments in the South, explained B.E. Walker: Nelles, *Politics of Development*, p.229.

p.39, Whitney sounded what must to the Toronto financiers have seemed: quoted in Denison, *People's Power*, p.46.

p.40, "We want no electric barons here . . .": Quoted in Nelles, *Politics of Development*, pp.272, 277.

p.40, It was a victory for Beck: quoted in Nelles, *Politics of Development*, p.291.

p.40, He denounced the government of the early 1920s: quoted in Denison, *People's Power*, pp.160, 136.

p.41, Planned "on a scale of the great European cathedrals": Denison, *People's Power*, p.127.

p.42, "He saw their agents under every gooseberry bush": E.C. Drury, *Farmer Premier: The Memoirs of E.C. Drury* (Toronto: McClelland and Stewart, 1966), p.136; financial data cited in Nelles, *Politics of Development*, p.415.

p.43, We have no way of knowing whether Beck: quoted in McKay, *Electric Empire*, p.23.

p.44, The year before the Depression hit, it added: Denison, *People's Power*, pp.185–86.

p.44, By 1926 Ontario's rate to domestic consumers: Denison, *People's Power*, pp.182–83.

p.45, According to Hydro historian Keith Fleming: Keith R. Fleming, *Power at Cost: Ontario Hydro and Rural Electrification, 1911–1958* (Montreal & Kingston: McGill-Queen's University Press, 1992), pp.16, 260.

p.45, It would not be until the mid-1950s that 96 per cent: Jean Christie, "Morris R. Cooke and Energy for America, in *Technology in America*, ed. Pursell.

p.45, "Public ownership in the electric power field": quoted in Nelles, *Politics of Development*, p.467; Nelles quote, same page.

p.46, The basic idea behind the promotional rate structure: quoted in Denison, *People's Power*, p.103.

p.47, The ads featured a smiling bride: see McKay, *Electric Empire*, esp. ch. 2 and p.45.

p.47, Lake found that the PUC was using high-frequency signals: Interview with Bob Lake, May 22, 2003.

p.49, "Hydro wants to control everything . . .": quoted in McKay, *Electric Empire*, p.222.

p.51, "Darlington is now the largest construction site in North America . . .": *Nuclear Free Press*, Summer 1985.

p.51, Dalton Camp noted that the Tories: quoted in Speirs, *Out of the Blue*, p.58.

p.53, "The cost of re-tubing Pickering A and Bruce A: *Nuclear Free Press*, Fall 1983.

p.53, It was predicting that it would need 30,000 MW: Howard Hampton, *Public Power: The Fight for Publicly Owned Electricity* (Toronto: Insomniac Press, 2003), p.132.

p.54, In the first week thirty-one people died from radiation exposure: Aleg Cherp et al., *The Human Consequences of the Chernobyl Nuclear Accident: A Strategy for Recovery*, United Nations Development Program and UNICEF, New York, Jan. 25, 2002; Nuclear Energy Agency, *Chernobyl: Assessment of Radiological and Health Impacts, 2002 Update of Chernobyl: Ten Years On*, Organization for Economic Cooperation and Development, 2002.

p.54, "Finishing Darlington was a renunciation of the basic economic principle . . .": Interview with David Poch, Oct. 10, 2003.

p.54, "I had a hell of a problem with Darlington: Interview with David Peterson, Oct. 9, 2003.

p.55, Scott later reckoned that Darlington's interest bill alone: Ian Scott, *To Make a Difference: A Memoir* (Toronto: Stoddart, 2001), p.145.

4 Liberalizing Electricity: Chile, Britain, and Ontario

p.57, Environmentalists were not the only ones manoeuvring: Richard Hirsh, *Power Loss: The Origins of Deregulation and Restructuring in the American Electric Utility System* (Cambridge, Mass.: The MIT Press, 1999).

p.58, The junta, as historian Eric Hobsbawn put it: Eric Hobsbawn, *Age of Extremes: The Short Twentieth Century* (London: Abacus, 1995), p.442.

p.58, . . . in the words of one of the ruling generals: quoted in José Piñera, "Chile," in *The Political Economy of Policy Reform*, ed. John Williamson (Washington: Institute for International Economy, 1994).

p.59, Some twenty years later the U.S. Department of Energy was still calling it: U.S. Department of Energy website <http://www.converger.com/eiacab/chile.htm> accessed Dec. 1, 2003.

p.59, According to Hernan Büüchi Buc: Hernan Büüchi Buc, "Reflections on Economic Reform in Chile," *Economic Reform Today* (Center for International Private Enterprise), vol.1 (1991).

p.59, The largest single privatization: Joseph Collins and John Lear, "Pinochet's Giveaway: Chile's Privatization Experience," *Multinational Monitor*, vol. 12, no. 5 (May 1991).

p.60, According to two Chilean economists: Ronald Fisher and Pablo Serra, "Regulating the Electricity Sector in Latin America," *Economia*, Fall 2000.

p.61, To the contrary, studies in the United States had found: Sharon Beder, *Power Play: The Fight to Control the World's Electricity* (New York: The New Press, 2003), p.88.

p.61, The rationale for pursuing the privatization enterprise: For details, see Beder, *Power Play*, ch.11.

p.62, Between 1988 and 2000, industrial consumers saw: Greg Palast, Jerrold Oppenheim, and Theo MacGregor, *Democracy and Regulation: How the Public Can Govern Essential Services* (London: Pluto Press: 2003), p.129.

p.62, National Power, one of the two privatized generation companies: Beder, *Power Play*, p.207.

p.63, There is strong evidence that manipulation: Office of Electricity Regulation (OFFER), *Pool Price: A Consultation*, February 1999, p.2.

p.64, "Part of Ontario Hydro's challenge," wrote Marion Fraser: Marion Fraser, "Demand Management in Ontario," *Public Utilities Fortnightly*, Nov. 1, 1992.

p.65, Between 1970 and 1998 Canada's Gross Domestic Product grew: Torrie Smith Associates, *Phasing out Nuclear Power in Canada: Toward Sustainable Electricity Futures*, Ottawa, 2003, p.8.

p.66, Well into the 1980s, the electric power industry: Torrie Smith Associates, *Phasing out Nuclear Power*, p.11.

p.66, The Plan was unequivocal: Ontario Hydro, *Providing the Balance of Power: Ontario Hydro's Plan to Serve Customers' Electricity Needs*, Toronto, 1989, p.7.

p.67, By 1989 Hydro had scaled back its estimates of growth: Ontario Hydro, *Providing the Balance of Power*, p.x.

p.67, But the DSP's authors pointed out: Ontario Hydro, *Providing the Balance of Power*, pp.7–8.

p.68, According to Rod Taylor: Interview with Rod Taylor, Nov. 5, 2003.

p.69, In the first three years of its modest DSM programming: "Ontario Hydro, Ontario Hydro's Corporate Agenda for the 1990s," notes for remarks by Marion Fraser, January 1993.

p.69, From the 1920s onwards the monopoly franchise: Walt Patterson, *Generating Change: Keeping the Lights On*, Sustainable Development Programme, Royal Institute for International Affairs, 2003, p.6.

p.70, "Uncertainty over demand is the main reason . . .": quoted in Amory Lovins, *Small Is Profitable: The Hidden Economic Benefits of Making Electrical Resources the Right Size* (Snowmass, Col.: Rocky Mountain Institute, August 2002), p.23.

p.71, In 1974 Schumacher had written that modest operations: E.F. Schumacher, *Small Is Beautiful: A Study of Economics as if People Mattered* (London: Abacus, 1974), pp.27–29.

p.71, According to Rod Taylor: Interview with Rod Taylor, Nov. 5, 2003.

p.72, Regarding it as his "fundamental obligation": Interview with David Peterson, Oct. 9, 2003.

5 The NDP Years: Clearing the Track for Privatization?

p.74, "Our party stands at a crossroads," Bob Rae wrote: Ontario New Democratic Party, *Greening the Party, Greening the Province: A Vision for the Ontario NDP*, Toronto, March 1990.

p.74, The utility had previously earmarked $240 million: Ontario Hydro, *Annual Report 1990*.

p.74, "I was never anti-nuke," he later said: Interview with Bob Rae, Nov. 7, 2003.

p.75, Rae's accidental government, whose electoral platform would later be described: Chuck Rachlis and David Wolfe, "An Insiders' View of the NDP Government of Ontario: The Politics of Permanent Opposition Meets the Economics of Permanent Recession," in *The Government and Politics of Ontario*, 5th ed., ed. G. White (Toronto: University of Toronto Press, 1997).

p.75, Provincial unemployment levels rose: Ontario Ministry of Treasury and Economics, "Budget Paper E: Ontario in the 1990s: Promoting Equitable Structural Change," *Ontario Budget: 1991*, Toronto, 1991.

p.77, Hydro historian Neil Freeman notes the irony: Neil Freeman, *The Politics of Power: Ontario Hydro and Its Government, 1906–1995* (Toronto: University of Toronto Press, 1996), p.172.

p.77, At first Brooks found it hard to comprehend: Interview with David Brooks, May 5, 2003.

p.78, Brian Charlton, who eventually did make it into the energy minister's chair: Interview with Brian Charlton, Jan. 10, 1998.

p.80, When Eliesen soon left to become the CEO of B.C. Hydro, Rae found himself: Interview with Bob Rae, Nov. 7, 2003.

p.80, For Rae it was "the smartest single appointment . . .": Bob Rae, *From Protest to Power: Personal Reflections on a Life in Politics* (Toronto: Viking, 1996), p.231.

p.80, According to Hydro insider Rod Taylor: Interview with Rod Taylor, Nov. 3, 2003.

p.81, . . . whom United Nations Secretary General Kofi Annan described: From Foreword, Maurice Strong, *Where on Earth Are We Going?* (New York: Alfred A. Knopf, 2000). Other material on Strong comes from Elaine Dewar, *Cloak of Green* (Halifax: James Lorimer and Company, 1995).

p.83, The premier was at least considering it: Thomas Walkom, *Rae Days: The Rise and Follies of the NDP* (Toronto: Key Porter Books, 1994), pp.253–54.

p.83, He informed Strong that there would be no privatization: Interview with Bob Rae, Nov. 7, 2003.

p.84, The Commission, chaired by Gro Harlem Brundtland: World Commission on Environment and Development, *Our Common Future* (Oxford: Oxford University Press, 1987), p.43.

p.84, In the words of one jaundiced observer: Lele, quoted in Timothy O'Riordan, "Ecotaxation and the Sustainability Transition," in *Ecotaxation*, ed. T. O'Riordan (New York: St. Martin's Press, 1997), p.8.

p.85, "Strong's meaning was clear," green journalist Tom Athanasiou observed: Tom Athanasious, *Divided Planet: The Ecology of Rich and Poor* (Boston: Little, Brown, 1996), pp.168–69.

p.85, The prophets of free-market environmentalism were urging: What was "believed to be the first criminal case involving air pollution credits" was reported in June 2004—with echoes of Enron (which was the biggest booster of emissions trading). See "Architect of Southern California Pollution Credit Program Accused of Fraud," Associated Press, Los Angeles, June 18, 2004.

p.85, "It was an incredible time": Interview with Allan Kupcis by Deborah Irvine, Feb. 28, 2002.

p.86, Strong cancelled Ontario Energy Board hearings: Strong, *Where on Earth Are We Going?* p.262.

p.86, The changes we're making provide a unique opportunity: Maurice Strong, *Where on Earth Are We Going?* p.266.

p.87, A 1992 study revealed that: David Argue, *The Economic and Environmental Implications of Fuel Switching* and *Addendum: Impact on Electricity Rates* (Toronto: Coalition of Environmental Groups for a Sustainable Future, January 1992).

p.87, Energy Minister Wildman noticed that Hydro's leadership: Interview with Bud Wildman, Jan.15, 1998.

p.87, Marion Fraser, an old Hydro hand who ran Hydro's conservation programs: Interview with Marion Fraser, Sept.18, 2003.

p.87, By 1991 Ontario Hydro was investing $179 million: Ontario Hydro, *Annual Report 1991*, pp.14–15.

p.88, Long after Strong had left the scene: Ontario Hydro, *Annual Report 1992*, p.9.

p.88, Hydro was forced to issue an update to its Demand Supply Plan: Ontario Hydro, *Annual Report 1992*, p.9; and Ontario Hydro, *1992 Update to Providing the Balance of Power: Ontario Hydro's Plan to Serve Customers' Electricity Needs*, pp.1–2.

p.89, Overall, Hydro's spending on environmental programs: Ontario Hydro, *Sustainable Development/Environmental Performance Report*, 1994; and Ontario Hydro, *Annual Report 1994*, p.10.

p.89, Fox was handed the "difficult task of overhauling . . .": Strong, *Where on Earth Are We Going?* p.267.

p.89, "Maurice came in and really significantly curtailed . . .": Interview with John Fox, Sept. 23, 2003.

p.90, "The recession and its impact on government spending . . .": Interview with Brian Charlton, Jan. 10, 1998.

p.91, Chairman Mo said that he was "delighted": Strong, *Where on Earth Are We Going?* p.263.

p.91, He even embarked on an hourly spot-market experiment: Strong, *Where on Earth Are We Going?* p.262.

p.93, Attracted by Farlinger's Tory ties: Marci McDonald, "Privatization Power Play," *Maclean's*, Feb. 12, 1996, p.30.

p.94, "Strong knew that an election was coming": Interview with Allan Kupcis by Deborah Irvine, Feb. 28, 2002.

p.94, "Ontario Hydro must prepare itself to compete . . .": W.A. Farlinger, Gordon Homer, and Brian Caine, *Ontario Hydro and the Electric Power Industry: Vision for a Competitive Industry*, Ontario Hydro, Toronto, June 22, 1995, p.15.

p.95, The arguments around privatization of Ontario Hydro: Farlinger, Homer, and Caine, *Ontario Hydro and the Electric Power Industry*, p. 22.

6 South of the Border: California, Enron, and the Yankee Alternative

p.97, In its first three years Efficiency Vermont reduced the need: Efficiency Vermont, *Year 2002 Annual Energy Savings Claim and Draft Annual Report*, April 2003, pp.4–5.

p.99, "The universe changed when major environmental groups . . .": Interview with Beth Sachs, Aug. 19, 2003.

p.100, "If using less electricity on the customer's side of the meter . . .": Interview with Blair Hamilton, Aug. 19, 2003.

p.101, "We're doing more than signing a new law": Quoted in Richard Hirsh, *Power Loss: The Origins of Deregulation and Restructuring in the American Electric Utility System* (Cambridge, Mass.: The MIT Press, 1999), p.246.

p.102, But, he added, the operators had not yet seriously considered: Walt Patterson, *Transforming Electricity: The Coming Generation of Change* (London: Royal Institute of International Affairs, 1999), p.121.

p.103, The free-marketer's bible, *The Economist*, referred to Kenneth Lay: Quoted in Sharon Beder, *Power Play: The Fight to Control the World's Electricity* (New York: The New Press, 2003).

p.103, Imagine a country-club dinner dance: Brian O'Reilly, "The Power Merchant," *Fortune*, April 2000.

p.104, To oppose Roosevelt's legislation the Edison Institute: Beder, *Power Play*, p.68.

p.104, In the land of free enterprise, government regulators capped: For details on how this system operated, see Greg Palast, Jerrold Oppenheim, and Theo

MacGregor, *Democracy and Regulation: How the Public Can Govern Essential Services* (London: Pluto Press, 2003).

p.105, According to a study by the California Public Utilities Commission: Michael Kahn and Loretta Lynch, "California's Electricity Options and Challenges: Report to Governor Gray Davis," California Public Utilities Commission, 2001, cited in Beder, *Power Play*, p.105.

p.105, And even during the ostensible shortages of 2000, the state was exporting: *Los Angeles Times*, Jan. 14, 2001, cited in Beder, *Power Play*, p.111.

p.106, Under the new rules, or lack thereof: Beder, *Power Play*, p.109.

p.106, The Ralph Nader consumer group Public Citizen pointed out: cited in Beder, *Power Play*, p.107, note 20.

p.106, A report to the state government by two deregulation supporters concluded: Paul Joskow and Edward Kahn, "A Quantitative Analysis of Pricing Behavior in California's Wholesale Electricity Market during Summer 2000," *The Energy Journal*, December 2002.

p.106, Later, in testimony before the U.S. Senate's Commerce Committee: Quotation from the testimony of S. David Freeman, chairman of the California Power Authority, before a subcommittee of the Senate Commerce Committee, cited in Richard Stevenson, "Enron Trading Gave Prices Artificial Lift, Panel Is Told," *The New York Times*, April 12, 2002.

p.107, Plants that had been closed for maintenance: Beder, *Power Play*, pp.107–8.

p.107, As R. Martin Chavez, a former head of risk management in energy trading: Joseph Kahn, "Californians Call Enron Documents the Smoking Gun," *The New York Times*, May 8, 2002.

p.108, Enron's CEO Jeffrey Skilling told an audience of electricity entrepreneurs: Quoted in Palast, Oppenheim, and MacGregor, *Democracy and Regulation*, p.134.

p.108, In 1997 an ideologist from the Heritage Foundation predicted: see Beder, *Power Play*, pp.144–5, 89, 91.

p.110, In the last two months of 2000 Enron made: David Barboza, "Despite Denial, Enron Papers Show Big Profit on Price Bets," *The New York Times*, Dec. 12, 2002.

p.110, The programs cut peak demand by 8.4 per cent: Devra Bachrach, Matt Ardema, and Alex Leupp, *Energy Efficiency Leadership in California: Preventing the Next Crisis*, Natural Resources Defense Council and Silicon Valley Manufacturing Group, New York, April 2003.

p.112, Setting up an efficiency utility like the one in Vermont: Interview with Chris Neme, Aug. 18, 2003.

p.112, Installing the little control units to regulate electricity flow: *Addison Independent* (Middlebury, Vt.), May 29, 2003.

p.113, When Sherry and Ron Machia of Sheldon decided to build a new barn: *The Burlington Free Press*, April 7, 2003.

p.113, "Business and the environment don't have to be at odds": *The North Star Monthly* (Danville, Vt.), March 2003.

p.114, U.S. renewable energy standards were, however, set: Tim Woolf, Geoff Keith, David White, and Frank Ackerman, "A Retrospective Review of FERC's Environmental Impact Statement on Open Transmission Access," background paper prepared for the Commission for Environmental Cooperation of North America, Montreal, Oct. 19, 2001.

p.114, In the case of California, spending on conservation programs: Bacharach, Ardema, and Leupp, *Energy Efficiency Leadership in California*.

p.114, Nationwide, the move towards electricity deregulation in the United States: Martin Kushler, Ed Vine, and Dan York, *Energy Efficiency and Electric System*

Reliability, American Council for an Energy Efficient Economy, Washington, D.C., April 2002.

p.115, Curtis Morin, president of an iron works, told the Public Service Board: *Barre Times-Argus*, Dec. 12, 2002, Jan.1, 2003.

p.115, A former schoolteacher with a graduate degree: Interview with Jane Whitmore, Aug. 19, 2003.

7 All Aboard the Privatization Express: The Harris Tories

p.118, In the years leading up to the Conservative election victory, Farlinger had helped Harris: Marci McDonald, "Privatization Power Play," *Maclean's*, Feb. 12, 1996, p.30; John Ibbitson, *Common Ground: Inside the Mike Harris Revolution* (Toronto: Prentice Hall Canada, 1997), p.53.

p.118, Farlinger, a career accountant specializing in corporate reorganization, admitted: Martin Mittelstaedt, "Farlinger New Chairman of Ontario Hydro: Toronto Businessman Says Privatization of Utility Is Prime Goal," *The Globe and Mail*, Nov. 8, 1995.

p.119, Power Workers' Union president John Murphy: Quoted in Martin Mittelstaedt, "Purge of Ontario Hydro Directors Draws Strong Criticism," *The Globe and Mail*, Jan. 12, 1996.

p.119, Clifford Maynes, an environmentalist member: Clifford Maynes, quoted in Ben Wolfe, "Privatization of Ontario Hydro," OEN *Network News*, vol. 7, no. 2 (November 1995).

p.120, Dave Martin of the Nuclear Awareness Project argued: Dave Martin, "Ontario Hydro: Privatization Is Not the Issue," OEN *Network News*, vol. 7, no. 3 (March 1996).

p.120, For Energy Probe's Tom Adams, privatization would offer: Tom Adams, "Privatizing Ontario Hydro to Achieve Environmental and Social Goals," speaking notes for presentation to "Plug into Ontario's Power Politics," Ontario Energy/Environment Caucus Annual Conference, April 9, 1994.

p.120, Journalist Thomas Walkom noted: Thomas Walkom, "Hydro Thorn Energy Probe Rooted on the Right: Pro-Privatization Empire Not Part of Environmentalism's Whole-Grain World," *The Toronto Star*, Aug. 23, 1997. By the mid-1990s the core group of Norm Rubin, Lawrence Solomon, and Tom Adams had been able to translate their brand of free-market environmentalism into a formidable organization with a $2 million per year budget. While often a source of suspicion or dismay to other environmentalists, Energy Probe came to its right-wing views in an honest way. Inspired by the same government-planning-is-bad, free-market-is-good analysis of the Austrian political philosopher Friedrich Hayek that politicians like Margaret Thatcher and Ronald Reagan put into practice, they argue that hidebound government planners are no match for the genius of the marketplace. But unlike most environmentalists who look to the state to help build "green" markets, Energy Probe looks to the private sector to lead the way, with government simply enforcing market-friendly laws such as tradeable pollution credits. The assumption is that market discipline will be a far more effective defender of the environment than any democratically elected government, even if multinational corporations dominate these markets.

p.121, According to Energy Probe founder Lawrence Solomon: Lawrence Solomon, "When It Comes to Regulation," editorial, *Next City* magazine <www.nextcity. com/main/town/2editor.htm>, posted March 12, 1996, accessed Dec. 9, 2003.

p.121, In one bit of biographical bravado, Energy Probe even described: From <www.urban-renaissance.org/urbanren/index.cfm?DSP=content&ContentID=547>,

accessed Dec. 12, 2003. The Urban Renaissance Institute is a project of the Energy Probe Foundation.

p.121, "There is a strong support for continued public majority ownership . . .": Ontario Hydro, *Competition, Convergence and Customer Choice: Submission to the Advisory Committee on Competition in Ontario's Electricity System*, Toronto, Jan. 25, 1996, pp.26, 29.

p.122, They were arguing that "there is no longer . . .": Ontario Hydro, *Competition, Convergence and Customer Choice*, p.11.

p.122, To make matters worse, the Power Workers' anti-privatization campaign: Howard Hampton, *Public Power: The Fight for Publicly Owned Electricity* (Toronto: Insomniac Press, 2003), p.192.

p.122, One insider pondered the problem of how to describe: Quoted in McDonald, "Privatization Power Play," p.30.

p.122, The corporate lawyer had mused: Martin Mittelstaedt, "Ontario Hydro Not Sacred Cow, Chairman Says: Committee Head Considers Privatization as He Reviews Debt-Laden Utility's Future," *The Globe and Mail*, Jan. 4, 1996.

p.122, "We recognized Ontario Hydro's historic importance . . .": Quoted in Richard Douglas, "Is Ontario Hydro For Sale? Provincial Commission Recommends Dissolving Utility's 90-Year Monopoly," *Electricity Today*, June 1996.

p.123, "The executive came to a decision early on . . .": Interview with Allan Kupcis, Feb. 28, 2002.

p.124, Maurice Strong confessed that the nuclear division: Paul Waldie, Janet McFarland, and Gayle MacDonald, "Power Failure," *The Globe and Mail*, Aug. 16, 1997.

p.124, Kupsis explained the decision that would cost him his job: Interview with Allan Kupcis, March 17, 2004.

p.125, The outside analysts found that Ontario's nuclear program: Paul Waldie and Chad Skelton, "Documents Itemize How Plants Earned Low Rating," *The Globe and Mail*, Aug. 15, 1997.

p.125, The report, issued in August 1997: Verbatim excerpts from Andognini report, published as "Trouble at Ontario Hydro," *The Globe and Mail*, Aug. 14, 1997.

p.125, Kupcis . . . took what he later described: Quoted in Jennifer Wells, "Nuclear Fallout," *The Toronto Star*, Sept. 27, 2003, p.D1.

p.126, According to Kupcis, the number one management priority: Interview with Allan Kupcis, Feb. 17, 2004.

p.126, "It appears that the chickens are coming home to roost": Quoted in James Rusk and Chad Skelton, "Hydro Fiasco to Cost Billions," *The Globe and Mail*, Aug. 14, 1997.

p.127, "The bottom line is that these proposals: Ontario Ministry of Energy, "White Paper Sets out Plan to Restructure Ontario Hydro," press release, Nov. 6, 1997.

p.128, The whole job—if the remaining three reactors could ever be made to work: Jake Epp, *Report of the Pickering 'A' Review Panel*, Toronto, December 2003.

p.129, According to Kupcis, when the big users: Interview with Allan Kupcis by Deborah Irvine, Feb. 28, 2002.

p.129, The evidence coming out of the United States and other deregulated jurisdictions: Greg Palast, Jerrold Oppenheim, and Theo MacGregor, *Democracy and Regulation: How the Public Can Govern Essential Services* (London: Pluto Press: 2003), ch.12.

p.129, If AMPCO members could use their muscle to grab long-term contracts: The prices are cited in Ontario Hydro, *Competition, Convergence and Customer Choice*, p.19.

p.129, Within a month of the Conservative Party's 1995 election victory: Association of Major Power Consumers of Ontario, AMPCO *Position on Electricity Market Restructuring*, Toronto, July 1995.

p.130, The chairman of the Stakeholders' Alliance was former MPP and Tory party president: See McFadden's law company profile at <http://www.gowlings.com/professionals/professional.asp?profid=720> and <http://www. gowlings.com/industry/infrastructure.asp> accessed June 16, 2004.

p.130, In its submission to the Macdonald Committee, AMPCO sounded a note: Association of Major Power Consumers of Ontario, AMPCO's *Final Submission to the Advisory Committee on Competition*, Toronto, March 10, 1996, pp.1–2.

p.130, Big industrial users such as Dofasco and Imperial Oil: Martin Mittelstaedt, "Heavy Hydro Users Seek Continued Rate Deal: Industrial Giants Ask Province to Extend Cheap Power Deals beyond Two Years or Risk 'Public Interest,'" *The Globe and Mail*, Jan.10, 2001.

p.130, "People are salivating all over the place": Peter Kuitenbrouwer, "The Privateers: Suitors Line up and Drool over Hydro," *Eye Magazine*, Feb. 22, 1996.

p.131, The competition-friendly Liberal Party: Fred Vallance-Jones, "Electric Power Firms Open Wallets for Tories," *The Toronto Star*, April 22, 2002.

p.133, "You don't have to have a Ph.D. in economics . . .": Interview with Bob Lake, May 22, 2003.

p.134, "There must have been at least a billion dollars spent: Interview with George Davies, April 24, 2003. The Acres Group is one of Canada's oldest and largest engineering and consulting companies, specializing in electric power planning and development.

p.134, Electricity is "deeply affected with the public interest": Mark Cooper, *All Pain, No Gain: Restructuring and Deregulation in the Interstate Electricity Market*, Consumers Federation of America, Washington, D.C., September 2002.

p.135, During the run-up to the commercialization of electricity in Ontario: Global Change Strategies International, *A Bird's Eye View of Electricity Supply and Demand to 2020: A Report Prepared for the Canadian Electrical Association*, Ottawa: July 23, 2001, pp.12–13.

p.135, "The government has failed on smog controls": Statement at the release of the *Green Report Card on Electricity Restructuring*, Feb. 26, 2001.

p.136, Tom Adams of Energy Probe, appointed to the committee: Personal communication. Adams also wrote, in a Jan. 11, 2002, posting to the on-line <cdn-nucl-l> discussion list: "Government subsidies to solar and wind power production (as distinct from research), whether in the form of net billing for transmission and distribution services, renewable portfolio standards (RPS), production tax credits, production bounties, non-fossil fuel obligations, or any other such schemes, should be cut off immediately. Such subsidies discourage success."

p.137, A quarter of the province's power was now being generated: Ontario Ministry of the Environment, *Coal-Fired Electricity Generation in Ontario*, Toronto, March 2001.

p.137, The OMA commissioned another study: Ontario Medical Association, *The Illness Cost of Air Pollution*, Toronto, June 2000.

p.139, The cost of production—about a penny per kilowatt-hour: Ontario Hydro, *Competition, Convergence and Customer Choice*, p.19.

p.140, (As it happened, in the eighteen months after the sale . . .): Martin Mittelstaedt, "Hydro Stations Sold for Less than U.S. Offer," *The Globe and Mail*, Feb.12, 2004.

p.140, Instead, The Tories dumped the proceeds: Paul McKay, "Eves Donor Cashed in on Power Deal," *The Ottawa Citizen*, Dec. 13, 2002.

p.141, "We sold power when Ontario needed it most: Quoted in McKay, "Eves Donor Cashed in on Power Deal."

p.142, "The opinion of financial houses internationally and domestically . . .": Interview with Jim Wilson, Aug. 8, 2001.

p.143, Interestingly enough, the loudest voice against the Bruce deal: Interview with Myron Gordon, June 16, 2001.

p.144, "Inevitably there is political sensitivity about opening the market: Interview with Andrew Johnson, July 11, 2001.

8   Danger Ahead: The Wheels Fall Off

p.145, What Harris did make clear: *The Globe and Mail*, Dec.13, 2001.

p.146, The outgoing premier sounded a reassuring note: CBC-Radio, May 1, 2003.

p.146, But right after Harris made his announcement: *The Ottawa Citizen*, Dec. 20, 2001.

p.147, "The bottom line is, we'll probably avoid . . .": *The Globe and Mail*, Jan. 18, 2001.

p.148, Together they concluded that fattening wallets: from Hydro One internal documents obtained by *Globe and Mail* reporter Martin Mittelstaedt in 2004; *The Globe and Mail*, March 23, 2004, March 24, 2004.

p.150, "In all my years of public life I have never witnessed . . .": Interview with Floyd Laughren, June 12, 2001.

p.150, . . . in 2000 the Harris Tories had ordered Ontario Power Generation: Jamie Swift, "A Shock to the System," ROB *Magazine*, February 2002; Martin Mittelstaedt, "Heavy Hydro Users Seek Continued Rate Deal: Industrial Giants Ask Province to Extend Cheap Power Deals beyond Two Years or Risk 'Public Interest,'" *The Globe and Mail*, Jan. 10, 2001.

p.151, Soon after the Hydro One sale was announced, Forstner's boss described: Swift, "Shock to the System"; *The Toronto Star*, Jan. 25, 2002.

p.151, Veteran Hydro-watcher Sean Conway: Quoted in Swift, "Shock to the System."

p.151, "Does it make sense to be paying the debt . . .": *The Toronto Star*, Dec. 4, 2001.

p.152, but prominent anti-nuke campaigner David Martin . . . immediately reacted: *The Toronto Star*, Dec. 4, 2001.

p.154, On a $5-billion deal the Royal Bank, Bank of Commerce, Toronto Dominion: *The Globe and Mail*, April 20, 2002.

p.154, Brascan, the company that would buy the money-spinning hydro dams: McKay, "Eves Donor Cashed in on Power Deal."

p.154, "Prices will go up and consumers will have little ability . . .": *The Globe and Mail*, Dec. 10, 2001.

p.155, "I don't see any alternative," AMPCO president Arthur Dickenson said: *The Toronto Star*, Dec. 15, 2001.

p.157, In Kingston, within a few weeks of the meeting: Kingston City Council, file no. CSU-C09-00-2002.

p.157, The success of the anti-privatization campaign was evident: Theresa Boyle, "Witmer Stuns Rivals over Hydro," *The Toronto Star*, Feb. 13, 2002; Ian Urquhart, "Debate Highlights Major Splits in Tory Party," *The Toronto Star*, Feb. 13, 2002.

p.158, "We'd seen the most horrible thing . . .": Interview with Paul Kahnert, June 17, 2003.

p.159, "They wanted to start with a blank sheet of paper": Interview with Charlene Mueller, June 17, 2003.

p.161, The green position of conservation and renewables: For details, see Keith Stewart, *Greening Public Power: Protecting the Public Interest in Electricity Restructuring*, Toronto Environmental Alliance, April 30, 2002.

p.162, Alberta had always been the odd man out: H.V. Nelles, "Hydro and After: The Canadian Experience with the Organization, Nationalization and Deregulation of Electrical Utilities," *Annales historiques de l'électricité*, a special issue on "Nationalisations et dénationalisations de l'électricité," no.1 (June 30, 2003), pp.117–32.

p.163, In the aptly named report ZAP!: RBC Dominion Securities, *ZAP! Alberta Is Jolted by Electric Deregulation*, Jan. 11, 2001.

p.163, The province's prices rose from 1.5 cents per kilowatt-hour: RBC Dominion Securities, *ZAP!*

p.163, Klein announced a billion dollars in electricity rebates: Government of Alberta, press release, Edmonton, Oct.11, 2000.

p.163, Meanwhile, the architect of the province's deregulation plan: Deborah Yeldin, "Alberta Thinks Better of Power Plan," *The Globe and Mail*, Nov. 7, 2003; Yeldin, "Power Deregulation Hits Albertans Where It Hurts," *The Globe and Mail*, May 12, 2003.

p.164, Renato Ruggiero, the first director general of the WTO: Renato Ruggiero, *Towards GATS 2000: A European Strategy*, address to the Conference on Trade in Services, Brussels, June 1998, quoted in Steven Shrybman, "A Legal Opinion Concerning the Impact of International Trade Disciplines on the Privatization and Restructuring of Ontario's Electricity Sector," Sack Goldblatt Mitchell, Toronto, December 2001.

p.164, All of which stuck in Shrybman's craw: Interview with Stephen Shrybman, Nov. 18, 2003.

p.166, British firms had succeeded in eliminating: House of Commons, *Debates*, Feb. 25, 1998, cited in Beder, *Power Play*, p.199

p.166, . . . many would have agreed with the assessment of Jim Stanford: Quoted in Paul Weinberg, "Pragmatist or Stooge? The Conversion of John Murphy from Labour Boss to Government Insider and Promoter of Ontario Hydro Sell-off," *Straight Goods* (on-line news service), Jan.15, 2001. We have relied on this analysis by Weinberg, who managed to get access to furtive PWU officials who did not make themselves available for interviews for this book.

p.166, So it was with John Murphy: *Straight Goods*, Jan.15, 2001.

p.167, "Eleanor Clitheroe," said one investment banker: *The Globe and Mail*, March 29, 2002.

p.168, "Right away there was this buzz . . .": Interview with Cecil Makowski, Dec. 14, 2003.

p.168, "'Hydro One,' the corporate name for the new millennium . . .": The Superior Court of Justice of Ontario, *Oral Judgment of the Honourable Mr. Justice Arthur Gans*, April 19, 2002, 02-CV-227522 CM 3.

p.170, *The National Post*'s Terence Corcoran warned: *The National Post*, April 20, 2002.

p.171, He declared that he would not be pressured: *The National Post*, May 2, 2002.

p.171, Within two years CEO Clitheroe's "compensation": *The Globe and Mail*, April 29, 2004; *The Globe and Mail*, June 13, 2002; *The Globe and Mail*, June 6, 2002.

p.171, Soon after that, when Eves confirmed: *The Globe and Mail*, June 15, 2002.

p.171, "What an awful situation," complained Energy Probe's Tom Adams: *The National Post*, May 31, 2002.

p.171, Hydro One board member Bernard Syron: *The Globe and Mail*, June 7, 2002; *The Globe and Mail*, July 3, 2002.

p.172, Dave Goulding, CEO at the new IMO, reported: *The Globe and Mail*, June 12, 2002.

p.173, Wilson replied, "The private sector asked us . . .": Interview with Jim Wilson, Aug. 10, 2001.

p.173, According to OPG vice-president Richard DiCerni: Interview with Richard DiCerni, Nov. 20, 2003.

p.175, As it stood with the sweetheart lease deal: Martin Mittelstaedt and Richard Mackie, "Nuclear Plant Leasing Deal Will Be Costly, Auditor Says," *The Globe and Mail*, June 7, 2002.

p.176, But the government, which had already exempted Hydro's successor companies: *The Globe and Mail*, June 22, 2003.

p.177, The Natural Resources report predicted: Natural Resources Canada, "Water Resources," in *Climate Change Impacts and Adaptation: A Canadian Perspective*, Ottawa, 2002 <www.adaptation.nrcan.gc.ca/> accessed Dec. 29, 2003.

p.178, This pollution rose from 28 megatonnes in 1990: *The Globe and Mail*, Nov. 7, 2002.

p.178, The price jolts on Ontario's new open power market were so dramatic: *The Globe and Mail*, Sept. 10, 2002, Oct. 19, 2002.

p.179, "If Canada is to achieve its climate change and clean air goals: Canadian Nuclear Association, "Canada's Climate Change and Clean Air Goals Need Nuclear Energy," press release, March 11, 2002.

p.179, John Baird, Ontario's third energy minister in as many years, confirmed: Alan Findlay, "Power Plant Re-Start Nuked?" *The Toronto Sun*, Sept. 10, 2002.

p.179, Taking the offensive, his report to the CNSC indicated: *The Globe and Mail*, Sept. 12, 2002.

p.179, When the British government privatized power, it found no takers: Beder, *Power Play*, p.202.

p.180, The market was increasingly dominated by a few powerful corporate hands: Beder, *Power Play*, pp.213–17.

p.180, The Power Workers' Union and the Society of Energy Professionals each doubled: Bruce Power, *The Power to Choose: Annual Review 2002*, Kincardine, Ont., 2003.

p.181, The utility had just paid the IMO $276 million for power: *The Globe and Mail*, Oct. 31, 2002.

p.183, Veteran Queen's Park watcher Graham Murray: Interview with Graham Murray, May 5, 2003.

p.183, According to Laval polar ecologist Warwick Vincent: *The New York Times*, Sept. 22, 2003.

9 Morbid Symptoms and a Regime Change

p.186, Iler recalled that the original idea of building an urban windmill: Interview with Brian Iler, May 6, 2003.

p.187, He came up with the theory of an "organic crisis": Antonio Gramsci, *Selections from the Prison Notebooks* (New York: International Publishers, 1971), p.276.

p.188, In the first year after the spot market was opened, the average price: Independent Electricity Market Operator, *The Ontario Wholesale Electricity Market: Year in Review 2002–2003,* 2003, p.4.

p.188, The terms of the market mitigation agreement required OPG: Ministry of Energy, *Backgrounder: Ontario Energy Board Amendment Act (Electricity Pricing), 2003 Highlights of the Changes,* Dec. 18, 2003.

p.191, In all, 263 power plants with 531 generating units capable of producing over 60,000 megawatts: U.S.-Canada Power System Outage Task Force, *Interim Report: Causes of the August 14 Blackout in the United States and Canada,* November 2003.

p.191, Of the eleven nuclear reactors that were operating at the time of the blackout: U.S.-Canada Power System Outage Task Force, *Interim Report,* p.92.

p.193, As David Cook of the North American Electricity Reliability Council (NERC) told the task force: From NERC's comments in response to the Interim Report of the U.S.-Canada Power System Outage Task Force, available at <http://www. electricity.doe.gov/documents/web_comments.pdf> accessed March 20, 2004.

p.193, North American companies got rid of more than 160,000 workers: UWUA's comments in response to the Interim Report of the U.S.-Canada Power System Outage Task Force, available at <http://www.electricity.doe.gov/documents/web_comments.pdf> accessed March 20, 2004; Greg Palast, Jerrold Oppenheim, and Theo MacGregor, *Democracy and Regulation: How the Public Can Govern Essential Services* (London: Pluto Press: 2003), p.134; Canadian Union of Public Employees (CUPE), *Deregulation, Privatization and the Ontario Power Failure,* CUPE Research Branch, September 2003.

p.193, It came as little surprise when the panel ignored evidence: John Wilson, "Notes for Presentation to Joint Task Force on the August 14 Blackout: Electricity Deregulation Caused the Blackout," Dec. 8, 2003.

p.193, First Energy CEO Anthony Alexander argued: From First Energy's comments in response to the Interim Report of the U.S.-Canada Power System Outage Task Force, available at <http://www.electricity.doe.gov/documents/web_comments.pdf> accessed March 20, 2004.

p.194, It came to light that within months of his appointment Wright himself: *The Globe and Mail,* Jan. 20, 2004.

p.194, Not surprisingly, Wright's task force found the government blameless: Wright was appointed on August 21, 2003, and finished his interviews by August 29. By August 26 he had given Eves a verbal report (which Eves used at press conferences) saying that Ontario was not to blame, although the written report wasn't delivered until March 14, 2004.

p.194, Watt was an attack-ad specialist whose consulting firm received: *The Globe and Mail,* Sept. 17, 2003.

p.196, McGuinty was quite clear in a mid-campaign newspaper article: *The Ottawa Citizen,* Sept. 19, 2003.

p.196, In the middle of the campaign, the greens stirred the pot: Jose Etcheverry, Stephen Hall, and Keith Stewart, *Bright Future: Avoiding Blackouts in Ontario,* David Suzuki Foundation, Vancouver, September 2003.

p.197, "Our plan will provide stable and predictable electricity prices . . .": Ministry of Energy, "Ontario Government Takes Responsible Action on Electricity Pricing," news release, Nov. 25, 2003.

p.197, In 1990 he had assured the nuclear industry that he was determined: Jake Epp, quoted in *Nucleonics Week,* Feb. 22, 1990, p.6; see "Jake Epp and Nuclear Power," Sierra Club of Canada backgrounder, June 3, 2003.

p.198, Epp helpfully offered his opinion that the cost: The Pickering A Review Panel, *Report of the Pickering "A" Review Panel,* December 2003.

p.198, The documents revealed that over five years: *The Globe and Mail*, Jan. 31, 2004.

p.199, All in all, top Tory insiders got $5.6 million in untendered Hydro One contracts: *The Globe and Mail*, Feb. 24, 2004.

p.199, The report opened by recognizing: Electricity Conservation and Supply Task Force, *Tough Choices: Addressing Ontario's Power Needs*, January 2004, pp.1, 3.

p.201, "The experience elsewhere," said Manley: *The Globe and Mail*, March 19, 2004.

p.202, The Manley report referred to this tendency politely: OPG Review Committee, *Transforming Ontario's Power Generation Company*, March 2004.

p.202, If anything were to go wrong, he predicted: *The Kingston Whig-Standard*, March 19, 2004.

p.203, In a much-anticipated speech to the Empire Club: Dwight Duncan, "Choosing What Works for a Change," speech to the Empire Club, April 15, 2004.

p.203, The future rested in creating "a climate that welcomes . . .": Duncan, "Choosing What Works for a Change."

p.205, A detailed analysis by a team of economists from Simon Fraser University: Mark Winfield, Matt Horne, Theresa McClenaghan, and Roger Peters, *Power for the Future: Towards a Sustainable Electricity System for Ontario*, Pembina Institute for Appropriate Development and the Canadian Environmental Law Association, May 2004.

p.206, As soon as he took office, Premier Whitney repeated a phrase: *The Globe* (Toronto), April 20, 1905, quoted in H.V. Nelles, *The Politics of Development: Forests, Mines and Hydro-Electric Power in Ontario, 1849–1941* (Toronto: Macmillan, 1974), p.257.

p.208, He came up with the idea of a "double movement" that operates: Karl Polanyi, *The Great Transformation: The Political and Economic Origins of Our Time* (Boston: Beacon Press, 1957), p.132.

p.209, Torrie's graphs showed that the productivity of Ontario's Candus: For a more detailed description of Torrie's plan, see Ralph D. Torrie and Richard Parfett, *Phasing out Nuclear Power in Canada: Toward Sustainable Electricity Futures*, prepared for the Campaign for Nuclear Phaseout, July 2003 <www.cnp.ca>.

Index

Acres Management Consulting 134
Adams, Tom 120, 136, 171, 225(n.)
Advisory Committee on Competition in
Ontario's Electricity System
(Macdonald Committee) 122, 126,
127, 130, 131, 147
air pollution 137–39
Alberta 162–64, 177
Alexander, Anthony 193
Allende, Salvador 57–58
alternative fuel sources 86–87, 98
Americans for Affordable Energy 102
Andognini, Carl 124–27
Annan, Kofi 81
anti-nuclear movement 4, 6, 8–14, 28,
31, 151
anti-privatization campaign 156–70, 173,
188
Association of Major Power Consumers
of Ontario (AMPCO) 6, 14, 28,
129–30, 144, 147, 150, 154
Athanasiou, Tom 85
Atomic Energy Control Board 10
Atomic Energy of Canada Ltd. (AECL) 6,
12–13, 16, 20, 68, 124, 197
automobile emissions 178

Babcock & Wilcox 9
Baird, John 179
Ballantyne, Morna 166–67
Bank of Commerce 36, 38
Bank of Montreal 154
Bank of Nova Scotia 154
B.C. Hydro 163
Beck, Adam 4, 14, 37, 38, 39–43, 45, 78,
118, 128, 131, 148–49, 151–52, 206
Beck II generating station 16, 139
Berlin Convention 37
Biddle, Tony 161
Big Blue Machine 4, 120
Birchbark Alliance, see Non-Nuclear
Network

blackout 189
August 2003 32, 108, 190–94
rolling 106–7, 110, 112
Board of Industrial Leadership and
Development (BILD) 30–31
Bothwell, Robert 16
Boudria, Don 51
Bourassa, Robert 72
Bradley, Jim 55
Brascan 134, 140–41, 154
Britain, see United Kingdom
British Energy plc 142–44, 149, 156, 174,
179–80
British Power Pool 101
Broadbent, Ed 8
Brooks, David 77–78, 126
Brown, Graham 198
Bruce generating station 1, 141–42, 156,
172, 174–75, 179–80, 184–85, 209
Bruce Power 142–44, 175, 179–80,
184–85
Buc, Hernan Büüchi 59
Budd, Peter 195, 200
Bush, George W. 102–3
Business Council on Sustainable
Development 84–85

California, deregulation in 101–16, 128,
134, 144, 147, 153, 156, 161, 163, 196
California Power Authority (CPA) 106–7
California Public Utilities Commission
101, 105
Cameco 180
Camp, Dalton 51
Campbell, Alex 24
Campbell, Ross 16
Campbell, Tom 51, 55
Campbellford PUC 49
Canada-U.S. Free Trade Agreement 56
Canadian Electrical Association 135
Canadian Energy Efficiency Alliance 89

Canadian General Electric (CGE) 12–13,
    16, 23, 34, 36, 47, 70
Canadian Imperial Bank of Commerce
    142, 154
Canadian International Development
    Agency 81
Canadian Nuclear Association 179, 202,
    210
Canadian Nuclear Safety Commission
    (CNSC) 179
Canadian Union of Public Employees
    (CUPE) 152, 158–61, 166–67, 180
    Local 1000 of 166–67
Canartech 56
Candu (Canadian deuterium uranium)
    reactors 13, 16, 52, 67, 84, 123, 124,
    136, 142, 175, 209–10
"CANTDU" 21
Cara Operations 172
Carr, Gary 182
Carter, Cyril 76–77
Carter, Jenny 76–77
Carter, Jimmy 28, 29
Cavanaugh, Ralph 196
Cavoukian, Anne 175
Centre for Policy Studies 60
Chalk River lab 20
Charlton, Brian 53, 76, 78–79, 90
Chavez, R. Martin 107
Chernobyl 53–54, 141
"Chicago boys" 58
Chile 57–60, 169, 207
Chrétien, Jean 177
Churchill River 72
Citizens for a Sound Economy (CSE) 102
Clark, Brad 182
Clarke, Bill 179
Clitheroe, Eleanor 82, 91, 94, 132, 147,
    150, 155, 167, 171–72, 194
coal/coal plants 6, 16, 19, 23, 25, 36, 62
    air pollution and 137–38, 173
    closure of 195, 202, 204–5, 206,
    209–10
Coalition for a Secure Energy Future 102
commercialization 5
Common Sense Revolution 118, 199,
    206
Communications, Energy and
    Paperworkers Union (CEP) 165–68
communism 30
competition/competitive market forces
    62–63, 91, 94–95, 122–23
Competition Bureau 153

conservation 27–28, 86–93, 100, 114,
    135–36, 161, 173, 195, 204
Conservation Law Foundation 98
Conservative Party (U.K.) 60
conserver vs. consumer society 22
Consumers First! 102
Conway, Sean 151
Cook, David 192–93
Cooper, Hugh 42
Copps, Sheila 51
Corcoran, Terence 170
Cordell, Arthur 25
Credit Suisse First Boston 154

Dadson, Aleck 134
Darcy, Judy 166–67
"Darlington Dozen" 21
Darlington generating station 64, 69,
    70, 78, 129, 205
    construction of 21, 24, 25, 27–28,
    30–31, 48, 51
    demonstration at 3–4, 8–14, 185,
    186, 208, 210
    Peterson government and 53–55
    Rae government and 74–76, 81–82,
    209
Darlington Information Centre 50–51
David Suzuki Foundation 196
Davies, George 134
Davis, William (Bill) 4, 14–15, 17–20,
    26–27, 30–31, 50–51, 75
Day, Sir Graham 147–48, 155
Day, Stockwell 148
"Death Star" 109
debt retirement charge 151, 169, 174
Decter, Michael 82
Deep River training centre 6
demand-side management (DSM) 27, 64,
    66–69, 71, 100, 135
Demand Supply Plan (DSP) 66–69, 79,
    88, 97, 200
    environmental assessment of
    68–69, 86, 90–91, 99, 120
Denmark 70
deregulation 59, 127, 153, 192–93, 207,
    see also privatization
    Alberta 162–64
    consequences of 107–10
    United States 101–16, 129, 150, 153,
    see also California
DiCerni, Richard 173
Dickenson, Arthur 155
Direct Energy 144

Dnieper River project 42
Dofasco 150–51, 154, 156
Doncaster, Deb 208
Douglas Point nuclear reactor 13, 18, 142
"Dream Team" 125–26
Drew, George 15
Drury, E.C. 42
Duncan, Dwight 197, 199, 203–5
Dundas, Peter 10

eco-efficiency 86, 88
economic liberalism 58, 60
Edison, Thomas 33
Edison Electric Institute 103–4
Edison General Electric 34
Efficiency Vermont 97–98, 111–16, 162
elections, provincial (1985) 51–52; (2003)
    194–97
Electric Development Company 36–39,
    140
Electrical Safety Authority 132
Electricité de France 180
electricity
    development of 32–56
    liberalizing of 57–72
    privatization of 57–62
    rates 5, 20, 27, 81–83, 92, 99, 136,
        156, 178, 182, 188, 197
Electricity Act (1998) 169
Electricity Competition Act 195
Electricity Conservation and Supply Task
    Force 195, 199
Electricity Transition Committee 166
Electric Power Research Institute 193
electrification 32–56
    rural 44
Eliesen, Marc 76–77, 79–80
Elliott, Brenda 118–19
embrittlement 52–53
ENDESA 59–60
Energy Competition Act 131, 139
"energy crisis" 4
energy efficiency 47–50, 64–73, 86,
    97–100, 111–16
Energy Efficiency Act 90
Energy Probe 28, 54, 64, 120–21, 134,
    136, 138, 200
Energy Star 115–16
Enron 5, 94, 101, 102–3, 105–10, 121, 131,
    134, 152–53, 156, 207
    collapse of 110, 153, 200
Environmental Assessment Act 25

environmental movement 22, 91–92,
    120–21, 135–39, 151–52, 160–61,
    173, 205
environmental protection 4, 6, 52,
    74–75, 79, 85–86, 195, 207
E.ON 180
Epp, Jake 197–98, 201
Espanola 88
Eves, Ernie 32, 154, 158, 168–71, 182–83,
    189, 191, 194, 196, 204
Exelon Generation Corporation 140

Fairley, Rob 160–61
Farlinger, William (Bill) 5, 93–94, 118,
    120, 122, 124, 138, 145, 198–99
Farmer-Labour coalition 40, 44
Fell, Tony 155
Ferguson, George Howard 44–45
First Energy 180, 192–93
Flavelle, Joseph 38
Fleming, Keith 45
Forstner, Gord 150–51
fossil-fuel plants 16
Foucault, André 165
Fox, John 89
Franklin, Robert 55, 64
Franklin, Ursula 23, 25
Fraser, Marion 64–65, 68, 87–88
Freeman, David 106–7
free market 57–63, 85, 91, 101, 106, 128,
    145–46, 181, 200, 204, 207
free trade 162, 164
Friedman, Milton 58
Frost, Leslie 15, 17
Frost, Robert 22, 117
fuel-cell technology 71

Galbraith, John Kenneth 15
"gaming" 63
Gans, Justice Arthur 168–69, 171, 174
gas turbines 71
General Agreement on Tariffs and Trade
    (GATT) 85
General Agreement on Trade in Services
    (GATS) 164
General Electric, see Canadian General
    Electric
Gerretsen, John 156–57
Gervan, Paul 11, 21
Gibbons, Jack 138, 208
Gillespie, Ed 102
Giuliani, Rudy 191
globalization 208

Goldman Sachs 142, 154
Gordon, Myron 143
Goulding, Dave 94, 172
Gramsci, Antonio 187
Grant, John 147, 149
Green Communities program 87, 89
greenhouse gases 84, 113, 137, 141,
    177–78
Greening the Party, Greening the Province
    74–76
Green Mountain Power 99, 111, 115
Greenpeace 11, 21, 161, 173
green power 74–75, 79, 114, 136, 195
Green Report Card 135–36
greens 6, 53, 54–55, 119–20, 187, 202,
    208–9
Grier, Ruth 53

Hamilton, Blair 98–100, 111, 116
Hampton, Howard 156, 161–62, 165, 195
"hard" vs. "soft" energy 22–23, 25–26,
    93
Harmer, W.R. 47
Harris, Mike 5, 93–94, 174, 182, 194,
    198–99, 204, 207
    market-opening and 145–51, 153–55,
    164, 169, 170
    privatization and 117–18, 120–21,
    126, 134, 135, 138, 144
Harrowsmith 24
Hawthorne, Duncan 179, 185, 193
Hayek, Friedrich 225(n.)
Hirsh, Richard 57
Hobsbawn, Eric 58
Hogg, T.H. 46
Holt, Al 79
Howe, C.D. 16
Hutton, Deb 117, 155, 194, 199
hydriding 52
Hydro-Electric Power Commission of
    Ontario 14, 17, 39, 40–47
hydroelectricity, development of 36–41
"Hydro-gate" 20
Hydro One 132, 143–44, 171–83, 194,
    198–99, 203
    anti-privatization campaign
    156–70, 173, 188
    cancellation of sale of 171–76, 183,
    188–90
    Initial Public Offering (IPO) 145,
    153–54, 167, 169
    privatization of 145–57
Hydro Québec 210

Iler, Brian 186
Independent Electricity Market Operator
    (IMO) 94, 116, 131, 132–34, 144,
    146–47, 153, 175, 176, 181, 189,
    200, 202
Independent Integrated Power
    Assessment (Andognini report)
    125–27, 143
independent power producers 78, 93
industrial revolution, Ontario 32–36
International Energy Agency 18
International Monetary Fund (IMF) 58,
    120
Investor Owned Utilities (IOUS) 29, 99

Jackson, Cam 181–82
Johnson, Andrew 144

Kahnert, Paul 151–53, 156–61, 165, 188,
    208
Kalkar 9
Karn, Gerry 194
Kearns, John 11
Kennedy, Thomas 15
Keynes, John Maynard 15
Keynesian economics 4, 17, 30, 58, 90,
    207–8
Klein, Ralph 163–64, 177
KPMG report, *see* John Manley
Kupcis, Allan 1–2, 3, 4, 85, 94, 123–25,
    129
Kyoto Accord 135, 152, 177–79

laissez-faire economics 5, 44, 58, 85,
    207–8
Lake, Bob 47–50, 65, 92, 133
Laughren, Floyd 2–3, 74, 83, 149, 151
Lay, Kenneth 102–3
least-cost planning 98–99
Leckie, Dan 186
Lewis, Stephen 80
Liberal Party 4, 51, 52, 71, 131, 157, 161,
    169, 182, 195–97, 201–5
liberalization 57–72
"Live Better Electrically" campaign 13,
    19, 27, 46, 47–48, 66, 206
load building 43–44, 47
load shaving 47–48
Long, Tom 93, 117, 120, 148, 199
Lovins, Amory 25–26, 70, 89, 161, 209
"low" technologies 25–26

Macaulay, Hugh 14
Macaulay, Leopold 14
Macaulay, Robert 14, 16–17
MacDonald, Donald C. 19, 27, 52,
Macdonald, Donald 122, *see also*
    Advisory Committee on
    Competition in Ontario's Electricity
    System
MacKinnon, Don 170, 194, 200
"Magnificent Seven," *see* Dream Team
Magrath, Charles 45
Makowski, Cecil 168
Manley, John 201–2, 204
Market Design Committee 131, 134, 136,
    147
market fundamentalism 4
market manipulation 63
Market Mitigation Agreement 131, 139
market-opening 146–57, 161, 170
Martin, David 120, 152, 208
Martin, Paul 81
Mayberry, John 154–55
Maynes, Clifford 119
McCallion, Hazel 138
McConnell, Lorne 68
McDermott, Dan 21, 208
McFadden, David 130, 193
McGrath, John 146
McGuinty, Dalton 157, 161, 195–96,
    202–5
McKay, Paul 10, 146
McKeough, Darcy 19, 27
McPhee, Andy 141
Meech Lake Accord 72
"Megawatt Laundering" 109
meters, electricity 47–48
MidAmerica 180
Miller, Frank 51–52
Ministry of Energy 19, 28
Moog, Gerhard 20
Morgan Stanley 131
Morrison, Neil 158
Mueller, Charlene 158–61, 208
Mulroney, Brian 56
Municipal Electrical Association 130
Murphy, John 119, 166
Murray, Graham 183

Nanticoke generating plant 16, 137
National Electric Light Association 104
National Energy Program (NEP) 30
National Grid Transco 164
National Power 62

National Resources Defence Council 98
National Trust 36
natural gas co-generation plants 78, 82,
    120, 205
Nautilus nuclear submarine 12, 13
Nelles, H.V. 38, 219(n.)
Neme, Chris 112
neo-liberalism 101
Nesbitt, Beatty 37
New Brunswick 24
New Democratic Party (NDP) 2–3, 4, 5,
    19, 52, 53–54, 72, 131, 161–62, 165,
    169, 182, 195, 197
    energy policy of 76–79
    government 73–95
    social contract 83
New Energy Directions 202
Newfoundland & Labrador 72
Niagara Falls, hydroelectricity and 34,
    35–37, 39, 40–41, 43, 45, 119, 139,
    204, 206
Nixon, Bob 55
Noble, Leslie 148, 199
Non-Nuclear Network 21
non-renewable fuel sources 16, 29
non-utility generators (NUGS) 29–30, 78,
    82, 92
North American Electricity Reliability
    Council (NERC) 192–93
North American Free Trade Agreement
    (NAFTA) 156, 164, 178
North Sea gas 62
NRG Energy Inc. 140
NRX 12
Nuclear Awareness Project 120, 208
nuclear power 5, 82, 124, 137, 140–41,
    197–98, 202, 205, 206, 209–10
    Bruce generating station 175, 179,
    181, 184–85
    dangers of 52
    Darlington generating station 9–13,
    27, 29
    moratorium on 90
    oil crisis and 18–19
Nuclear Power Demonstration (NPD)
    plant, *see* Rolphton

Office of Energy Conservation 25, 77
oil crisis 18–19, 98
Onstein, Peter 10
Ontario Clean Air Alliance (OCAA) 138,
    173, 208

Ontario Coalition for Nuclear
Responsibility 21
Ontario Electricity Coalition (OEC) 151,
152, 156, 157, 158, 160–61, 165–66,
168, 173, 188, 208
Ontario Electricity Financial Corporation
132, 140–41, 174, 189
Ontario Energy Board (OEB) 19, 86, 91,
123, 149, 153, 203
Ontario Hydro 149, 151, 187, 206–7, *see
also* Hydro-Electric Power
Commission of Ontario
accountability of 55–56, 74, 127
Andognini report 125–27, 143
board of directors 79–81, 118–19,
125–26
compared to Vermont IOUS 99
conservation programs 86–93
Darlington protest 3–4, 8–14, 185,
186, 208, 210
debt of 82, 132, 140–41, 151, 169,
174
Deep River training centre 6
demand-side management 27, 64,
66–69, 71, 88, 97
Demand Supply Plan (DSP) 66–69,
79, 88, 97, 200
Energy Management Branch 64
energy efficiency plans 64–72, 74,
86–93
environmental assessment of DSP
68–69, 86, 90–91, 99, 120
environmental movement and 22,
120–21, 135–39
expansion of 19, 27
Financial Restructuring Committee
93–94
gigantism of 64, 66, 70
inquiries into 26–27, 50, *see also*
Porter Commission
Rae government and 73–95
origins of 14, 38
Pickering nuclear accident 52–53
postwar period and 15–16
privatization of 2–3, 5, 64, 72, 80,
83, 91, 93, 117–44, 145
public vs. private ownership 78–79,
82–83
select committee hearings on
26–27
unbundling of 121–44, 145
*Ontario Hydro and the Electric Power
Industry* 94–95

Ontario Municipal Employees
Retirement Board (OMERS) 180
Ontario Power Authority 204
Ontario Power Company 42
Ontario Power Generation (OPG) 148,
150, 154, 166–67, 169, 173–75, 181,
198, 200–3
Bruce Power and 185, 188–89
creation of 131, 133, 136–39, 140,
142–43
Ontario Sustainable Energy Association
(OSEA) 208
Organization of Petroleum Exporting
Countries (OPEC) 18–19, 24
Osborne, Ron 198
Otonabee River 33–34, 37, 48

Patterson, Walt 69–70, 102
Payne, Brian 165
peak load 43
Peckford, Brian 72
Pellatt, Henry 39–40
Pennsylvania 36
Peterborough 33–35, 37, 47–50
Public Utilities Commission (PUC)
47–50, 65, 92, 119, 133
Peters, Eric 173–75
Peterson, David 27, 51–55, 65, 71–72,
199
Peterson, Jerry 70
photovoltaic panels 70–71
Pickering 9, 13, 16, 18
generating station A 1–2, 18, 48,
128, 172, 175, 179, 188, 191,
197–98, 201, 204–5, 209
nuclear accident 52–53
review panel on 198
Pinochet, Augusto 57–58, 60, 70
Poch, David 54, 135
Polanyi, Karl 207–8
Porter, Arthur 20
Porter Commission 20, 21, 26, 28, 50,
54, 65
postwar period 15–16
Power Corporation 81
Power Corporation Act 55–56, 77, 86
Power Workers' Union 48, 151–52, 180,
194
anti-privatization and 159–60,
166–68
privatization and 119–20, 122, 123,
130
Prince Edward Island 24

private power generation 78, 93
privatization 57–62, 64, 73–95, 130, 167,
    195, 207, 208
    of Hydro One 145–57
    of Ontario Hydro 2–3, 5, 64, 72,
    80, 83, 91, 93, 117–44, 145
Progressive Conservative party 4, 5,
    26–27, 30, 38–39, 52, 71, 161, 194,
    195, 197, 198–99, 206
    Davis government 14–19
    Harris government 117–44
    Red Tory wing 15
promotional rate structure 46, 47
Public Citizen 106
public ownership 3
public participation 23–24, 86, 91
Public Power 162, 195
public utilities commissions (pucs)
    47–50, 92, 132–33
Public Utilities Regulatory Policies Act
    (purpa) (U.S.) 29, 92

Quebec 72, 98, 199
Queenston-Chippewa project 41–43

Rachlis, Chuck 75
Rae, Bob 51, 73–76, 79–80, 83, 89,
    93–94
rates, electricity 5, 20, 27, 81–83, 92, 99,
    136, 156, 178, 182, 188, 197
Reagan, Ronald 30, 58, 101, 225(n.)
regulation 3, 19, 29
Reliant Resources 164
Remembrance Day price freeze 182, 187,
    189, 196–97
renewable energy 186–87, 204–5
Renewable Portfolio Standards (rps) 113,
    135–36, 227(n.)
re-tubing of reactors 53, 120, 204
Rhodes, Paul 148, 194, 199
Richardson Greenshields 131
Rickover, Hyman 12
"Ricochet" 109
Rio Earth Summit 81, 84–85
Robarts, John 15, 17
rolling blackouts 106–7, 110, 112
Rolphton Nuclear Power Demonstration
    (npd) plant 9, 12, 13, 17
Roosevelt, Franklin D. 104
Ross, George William 38–39
round-tripping 109
Royal Bank of Canada 131, 154
Rubin, Norm 225(n.)

Ruggiero, Renato 164
rwe 180

Sachs, Beth 98–99, 116
St. Lawrence Seaway 16, 46
"Savings by Design" program 69
Schumacher, E.F. 71
Science Council of Canada 23–26
Scott, Ian 55, 90, 198
Select Committee on Energy 53
Shrybman, Steve 162, 164–65, 167
Sierra Club 208
Silano, Bruno 159–60
Skilling, Jeffrey 108, 152, 193
Snobelen, John 123
snug buildings 3, 4
Society of Energy Professionals 180
"soft energy path" 4, 6, 22–23, 25–26,
    29, 70–71, 77, 92, 93, 99, 205, 210
solar power 23, 56, 98, 207, 227(n.)
Solomon, Lawrence 121, 225(n.)
South America 37–38
Southern California Edison 102
Speirs, Rosemary 31
Stakeholders' Alliance for Electricity
    Competition and Customer Choice
    129–30, 154
Stanford, Jim 166
Stockwell, Chris 158
Strange, Susan 108
Strauss, Lewis 18
Strong, Maurice 5, 80–94, 118, 124, 202
Sun Day 21, 22, 68
sustainability 3–4, 29
sustainable development 84, 86, 88
Syron, Bernard 171–72
Systems Benefit Charge 114

Taylor, Rod 68, 71–72, 80, 130
Tennessee Valley Authority 92, 107
Thatcher, Margaret 30, 57, 60–62, 70,
    71, 101, 119, 121, 207, 225(n.)
Three Mile Island 9, 10, 20, 25, 28, 141
Toogood, John 148
Toronto Dominion Bank 154
Toronto Electric Light Company 36
Toronto Environmental Alliance 152, 161
Toronto Hydro 132, 158–59, 181, 210
Toronto Renewable Energy Co-operative
    (trec) 186–87, 208
Toronto Street Railway Company 36
Torrie, Ralph 6–7, 21–23, 25, 65–66,
    161, 209–10

"traction boom," South American 37–38
TransAlta 155
TransCanada Pipelines 180
Trudeau, Pierre 25
Twenty-First Century Energy Project 102

"unbundling" 121, 131
United Kingdom 128, 132, 134, 155, 166,
    207
    British Energy 179–80
    electricity privatization 57, 60–62,
    83, 101–2
United Nations 84–85
United States 44, 59, 61, 62, 68, 92
    deregulation in 96–16, 129, 150,
    153
    energy policy in 28, 29, 30–31, 154,
    162, 178, 191
    nuclear power in 12, 18
uranium 6, 19, 23
Urban Renaissance Institute 226(n.)
U.S.-Canada Power System Outage Task
    Force 192
Utility Workers Union of America 193

VendingMiser devices 112
Vermont 96–101, 111–16
Vermont Energy Investment
    Corporation (VEIC) 97–98, 112,
    115–16
Vermont Public Service Board 100–1,
    111, 115
Vincent, Warwick 183
Vise, Joe 3–4, 8–14, 25, 31, 210

Walker, B.E. 38
Walkerton 6
Walkom, Thomas 120
Wallace, Henry 33
Washington Electric Co-operative 100,
    111
Watt, Jamie 194
weather, electricity demand and 172,
    176–78
Weinberg, Paul 229(n.)
welfare state 15
West, Steve 163–64
Whitmore, Jane 115–16
Whitney, James 38–39, 206
Wightman, Donald 193
Wildman, Bud 75, 83, 87
Wilson, Fred 165
Wilson, Jim 127, 136, 142, 166, 173

Wilson, Pete 101–2
wind power 23–24, 56, 70, 98, 185–87,
    204, 207, 208, 210, 227(n.)
Witmer, Elizabeth 157–58, 170
Wolfe, David 75
World Bank 120
World Commission on Environment
    and Development 84
World Trade Organization 85, 164
Wright, Glen 194